D0409994

Maps and Politics

PICTURING HISTORY

Series Editors
Peter Burke Sander L. Gilman
Roy Porter Bob Scribner

In the same series

Health and Illness
Images of Difference
Sander L. Gilman

The Devil
A Mask without a Face
Luther Link

Reading Iconotexts
From Swift to the French Revolution
Peter Wagner

Men in Black
John Harvey

Eyes of Love
The Gaze in English and French Paintings
and Novels 1840–1900
Stephen Kern

The Destruction of Art
Iconoclasm and Vandalism since
the French Revolution
Dario Gamboni

The Feminine Ideal
Marianne Thesander

Trading Territories
Mapping the Early Modern World
Jerry Brotton

Picturing Empire
Photography and the Visualization of the British Empire
James R. Ryan

Pictures and Visuality in Early Modern China
Craig Clunas

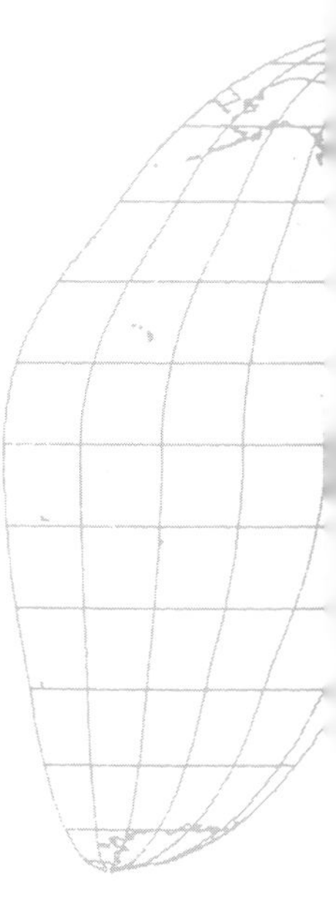

Maps and Politics

Jeremy Black

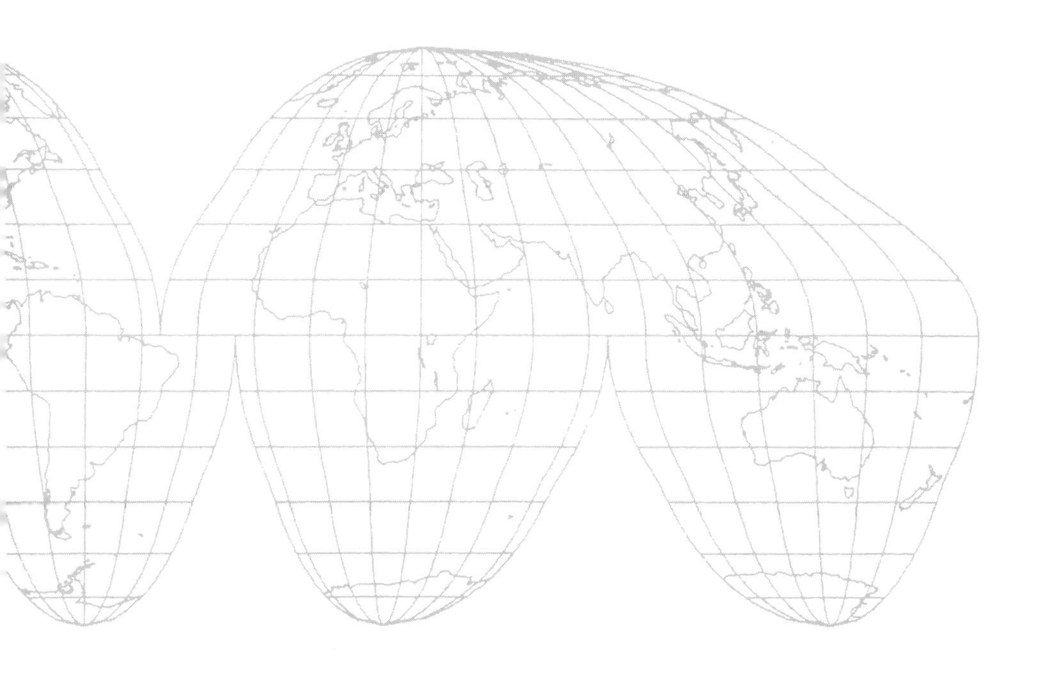

REAKTION BOOKS

For Roy Porter

Published by Reaktion Books Ltd
11 Rathbone Place, London W1P 1DE, UK

First published 1997

Jacket designed by Ron Costley
Designed by Humphrey Stone

Photoset by Wilmaset Ltd, Wirral
Colour printed by BAS Printers, Hants
Printed and bound in Great Britain by Biddles Ltd,
Guildford and King's Lynn

British Library Cataloguing in Publication Data:
Black, Jeremy,
Maps and politics. – (Picturing history)
1. Historical geography 2. Cartography – Political aspects
3. Maps – Political aspects 4. Territory, National
5. Boundaries
I. Title
911'.09
ISBN 1 86189 012 5

Contents

RAILWAY MAP

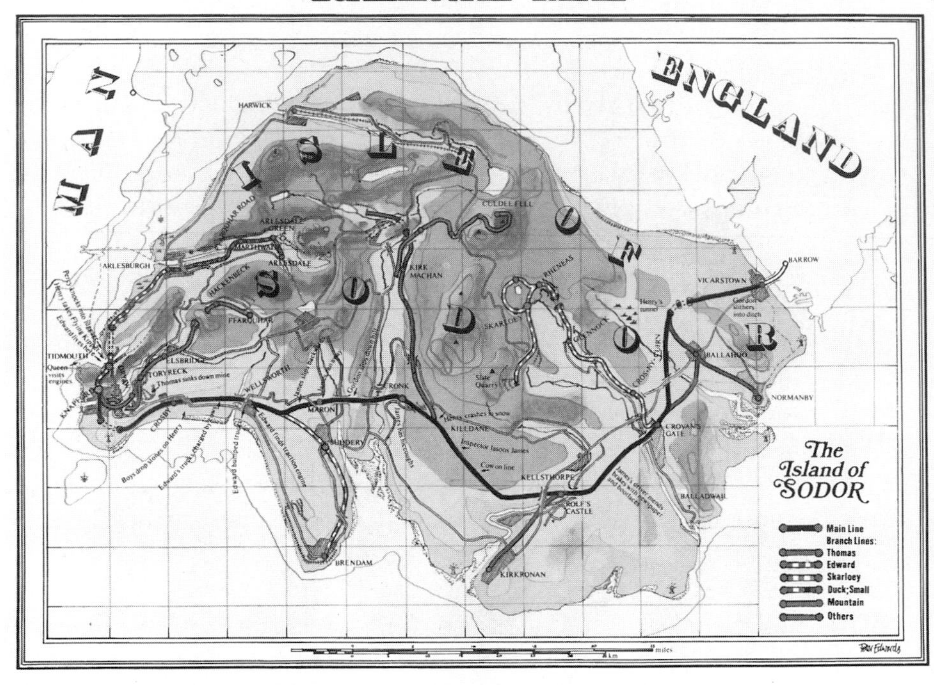

The detailed map of an imaginary land that is more real to many children than the maps of non-fictional countries. The island of Sodor from the Reverend W. Awdry's *Favourite Thomas the Tank Engine Stories* (1991). The doings of Thomas the Tank Engine and other friends can be readily followed. This is a world in which travelling is all, and the routes of journeys are of far less importance.

Preface

Books with maps in them are among my early memories of reading, the imaginary maps that Arthur Ransome drew for his saga of childhood adventure, *Swallows and Amazons*, and the maps of medieval battles and campaigns produced by R.R. Sellman. As a child I drew maps of an imaginary country, its tactility, indeed reality, created by mapping. At school I took both geography and history to 'A' level, and the choice of which subject to follow at university level was difficult. Concerned about the apparent mathematicization of geography, I chose history, and did not pursue maps: historical geography 'belonged' to the geographers. I remained interested in maps, but a scholarly concern did not return until 1992. Since then I have increasingly worked on historical cartography and related topics. This has led to two projects. *Maps and History*, a lengthy work tracing historical cartography from its origins and concentrating on historical atlases, is to be published in 1997. *Maps and Politics*, the present book, is designed as a shorter study, more closely focused on recent and contemporary concerns. It is linked to a course I am introducing at the University of Exeter.

I have benefited greatly from the friendliness of the map world. I would like to thank in particular the exemplary staff of the Map Room of the British Library. I have also benefited from work in the map rooms of the National Library of Scotland and the libraries of Ball State, British Columbia, Cambridge, Colorado (Boulder), Denver, Newcastle, Oxford, Texas, Texas Christian, Western Ontario and York (Ontario) Universities. I have benefited from the advice of John Andrews, Paul Harvey, Robert Peberdy, Charles Withers and three anonymous readers on earlier versions of the work. I am grateful for the opportunity to develop some of the following ideas in lectures at the University of Illinois (Urbana), the Warburg Institute and Guildhall University. I owe much to the secretarial support of Wendy Duery.

This book is dedicated to a historian I greatly admire. Roy Porter is not only a superb scholar with an amazing range and an accessible style, he is also all too rare among modern, academic historians in being able and willing to make the past immediate and interesting to the wider public. That Roy is an engaging, lively and humane individual is also worth mentioning.

Introduction

Maps have played and play a major role in politics, both international and domestic, reflecting the powerful ability of visual images and messages to represent and advance agendas. The major development of maps in this field over the past century has been a response both to cartographic advances and to a greater emphasis on graphic imagery in societies affected by politicization, democratization, and consumer and cultural shifts that emphasize the visual.

That many maps treat of politics is readily apparent; this was true from the outset of cartography. There was a close connection in the ancient world between map-making and imperial conquest and rule, between what purported to be world maps and pretensions to world power.[1] An emphasis on political power remains true of much modern cartography: maps are used both to assert territorial claims and to settle them, especially frontier disputes; and political preferences at elections are often presented in terms of maps.

Yet most maps and mapping do not seem to bear any reference to politics to the map-purchaser and user, whose principal access to cartography is when he or she is lost and searching for the correct or best route, or bored and flicking through an airline magazine that depicts on its maps the apparent routes of the company's aeroplanes: the world reduced to order and spanned by humanity. Most purchasers and users see the development of map-making as a science based on changes in mathematics, perspective and surveying. In our time this impressive trajectory continues, with the availability of satellite surveying, computer-processing of data, sophisticated colour printing and other developments. The consequences are heady. Maps can be produced faster and more plentifully than ever before. They can be recentred readily, and different projections and perspectives can easily be adopted. Complex and extensive data sets can be rapidly mapped. The invariable cartographic characteristics produced by the use of coordinates have now been joined to the computer processing and depiction of data to create a cartographic technology that appears to offer scientific precision and comprehensiveness.

Most users rely on the apparent accuracy and objectivity of maps; they do not see the very process of mapping as political. Is this correct? Can politics be treated as a sub-set of cartography, a matter of subject specialization and/or readily apparent bias, but one that is separable from the vast bulk and purpose of cartographic production and use? Or are the power and purpose of maps inherently political? This book addresses these important questions and seeks to emphasize that the apparent 'objectivity' of the map-making and map-using processes cannot be divorced from aspects of the politics of representation.

This enquiry is of contemporary importance because the role of place and space is of importance, and indeed of growing importance, in a number of disciplines. Just as time, as subject and explanation, does not 'belong' to historians, so too with space and geography. Anthropologists, historians and sociologists[2] are among those concerned with the multiple and contested roles of space. Maps do more than record such interests, because mapping is central to attempts to advance, record and contest understandings of space and spatiality.

1 Cartography as Power

Maps are selective representations of reality; they have to be. Even if maps were to be life-size photographs they would be distortions: a three-dimensional, spherical object, such as the globe, cannot be presented in two dimensions without its essence being altered, and this problem affects the mapping of parts of the globe. Yet, even if that problem could be overcome, and life-size, photographic, maps be produced in some futuristic virtual-reality technology, there would still be the question of how the photography/cartography should be presented. What perspective would be employed? Would there be shadows? Would the map be in darkness and, if not, why not? A landscape in twilight or darkness, the human presence, indeed power, etched in light, is as 'realistic' as the total vision of unclouded daylight, indeed more so, and yet this vision is employed in most maps (the most prominent exceptions being weather maps, which often include cloud). Furthermore, a 'daylight' map, whether life-size or not, with its misleading simultaneity of perception, is not affected by the methods by which humans on the ground, in 'real life', seek to create perceptions, to use light and lights to define space and create, or prevent, perceptions.

Maps are not life-size; they are models, not portraits, let alone photographs, of life. Most are minute compared with what they depict. As a result, map-makers have to choose what to show and how to show it, and, by extension, what not to show. The word 'show' is deliberately chosen. It conveys a sense of art and artifice, of the map-maker as creator rather than reflector. A map is a show, a representation. What is shown is real, but that does not imply any completeness or entail any absence of choice in selection and representation.

Some of the most striking modern images of the Earth are produced by orbiting satellites. Since the 1970s, NASA, the American National Aeronautics and Space Administration, has been producing such pictures, by a technique known as Remote Sensing by Landsat Imagery, which generates images from electro-magnetic radiations outside the normal visual range. These might appear natural maps, the product of human artifice, but essentially maps as photographs, photographs as maps. However, human intervention in the creation of the image is more direct than many viewers appreciate. For example, the use of different wave lengths leads to a stress on various aspects of the Earth's surface. Infra-red is especially valuable for vegetation surveys and for water resources. Choice and context greatly affects the impression that is created. In particular, because the image and colour-range are unfamiliar to most viewers, the caption and accompanying text are especially important in influencing the responses of viewers.

A map is no mere illumination. A minuscule-scaled photograph from a high-flying aeroplane of, say, Eastern Ontario would be too crowded for it to be possible to discern much, other than a concentration of human activity in the form of light near the lake. As the aeroplane flew lower, more would be revealed in photographs, but a map is not a photograph. The choice of what to depict is linked to, and in a dynamic relationship with, issues of scale *and* purpose, and the latter issue is crucial. A map is designed to show certain points and relationships, and, in doing so, creates space and spaces in the perception of the map-user and thus illustrates themes of power. This is readily apparent with two very common types of map, which are produced at very different scales.

The first, the map of the world, or a region thereof, divides up its land space (although not generally the seas) in terms of territorial control and political authority: the map as assertion of sovereignty. States, such as France and Germany, are the building blocks of such a map. As will be discussed, other methods of organizing space at this scale, indeed of presenting political space, are ignored.

Such mapping, using states as building blocks, does not have to be explicitly political. For example, weather maps are one of the most familiar types of maps. They might seem totally removed from the world of politics. However, their frame of reference is generally that of a political unit, say, Italy or Spain, in part because they commonly appear in national newspapers or on national television or radio, or because such states are the most convenient units for discussion and depiction. Thus, in Britain, an inhabitant of Kent is provided with more information about the situation in distant Westmorland than in nearby Pas-de-Calais, which is in a different nation-state. The former is assumed to be more relevant, but that is not the essential reason for the scope of the weather map. Instead, it is a statement of the centrality of the national sphere even in fields in which the state, indeed the country, plays no role apart from the provision of the weather service.

The second very common type of map is that of a city or town, or a detailed part thereof. These maps are organized around streets; indeed, that is how they are indexed when produced together in street atlases, such as the London *A–Z*. The index plays a crucial role in such atlases. Railways are marked, but they are shown as thin lines and are not indexed. It is only on the imaginary island of Sodor, the scene of the 'Thomas the Tank Engine' stories, that a modern British cartographer can map roads alongside railways and make the former far less prominent.

In the 'A–Z'-ing of life, habitations emerge as the spaces between streets. Differences within the city or town, for example of wealth, or environmental or housing quality, are ignored. The perceptions that create and reflect senses of urban space, often rival, contested and atavistic, are neglected, in favour of a bland uniform background that is described, and thus explained, insofar as there is any explanation, in terms of roads.

This is not a world of neighbourhoods, of upwardly mobile or downwardly mobile quarters, of areas largely inhabited by families or singles. Such blandness is necessary in order to highlight the roads, although, in addition, the network the roads thus display is misleading, because the detailed maps generally do not accept a hierarchy of routes, or, if they do so, do not give it sufficient prominence. A road map of a city does not depict roads in proportion to their traffic density, although it is increasingly common to distinguish main streets by colouring their surfaces. In addition, most city maps show major public buildings (typically in black), and thus allow city centres to be distinguished in a rough and ready way.

However, maps of cities are very much ground level. There is little, if any, suggestion of the vertical, and thus of the many who live and work in skyscrapers or more modest multi-storeyed buildings, and indeed of the transport problems these buildings pose and the links they offer (stairs/lifts). This elision of the vertical dimension of urban life is an aspect of the emphasis on ground-transport routes, particularly roads. The city is a space to be traversed, a region to be manipulated or overcome in the individual's search for a given destination, not an area to be lived in and through. Far from being composed of neighbourhoods, the city is a sphere of distance to be negotiated, indeed overcome, by road. More generally, the structure, typology and density of activity in the city is neglected.[1]

An *A–Z* road map does, of course, provide some hints about urban environmental quality. The presence of parkland is one important indicator. On the *Nicholson Colour London Streetfinder* (London, 1985) I use, there is an obvious contrast between the green spaces and graphic openness of the Hampstead area on p. 46 and Brixton (p. 90), where no green spaces are shown. They provide a hint of urban character for unvisited areas, and reinforce the perceptions already formed of parts of the town that have been visited.

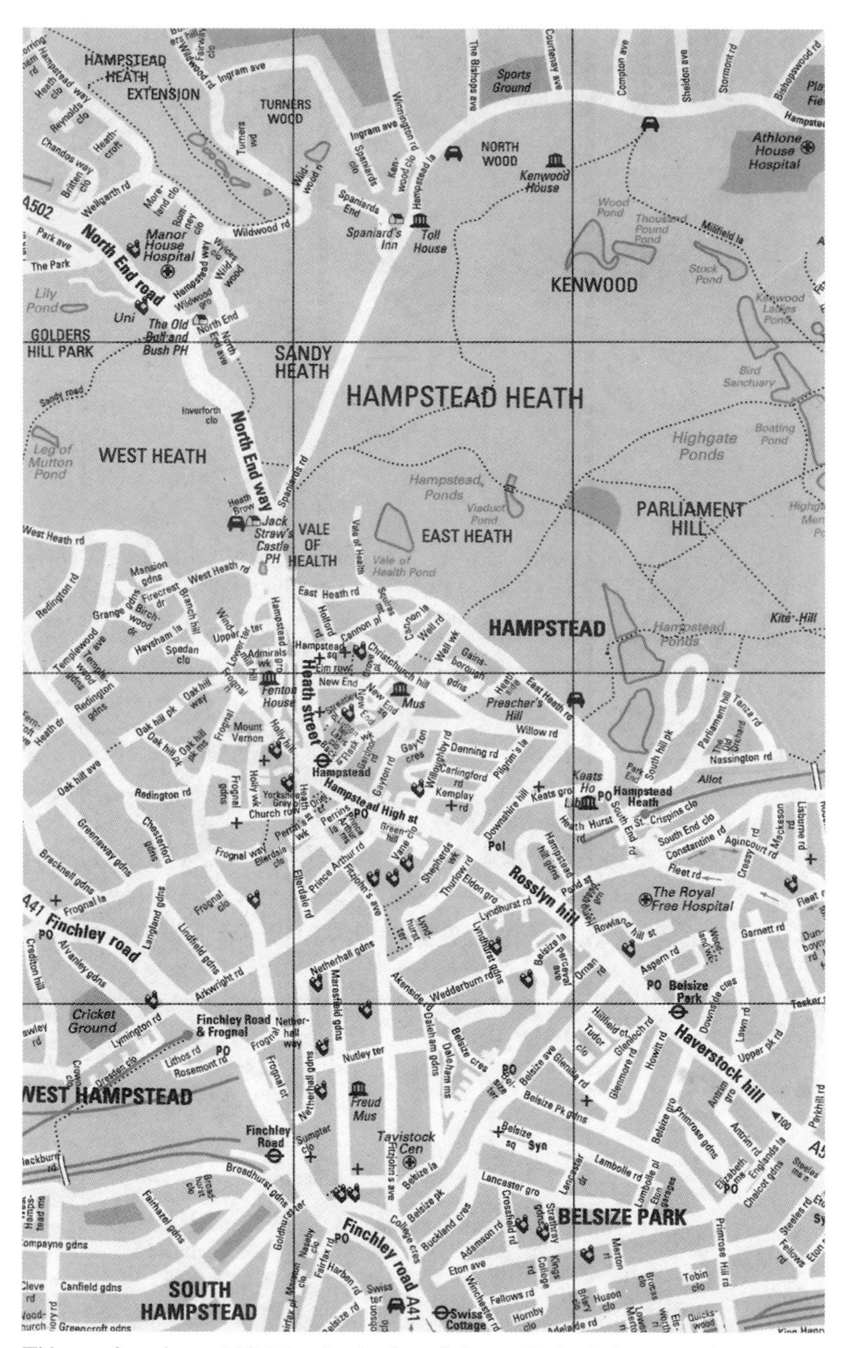

This map from the 1996 *Nicholson London Streetfinder* reveals clearly the contrast between Hampstead, with its plentiful green spaces, and neighbouring areas, such as Kentish Town, that lack them. Other amenities, for example the views from the Heath and the way in which Hampstead literally looks down on its neighbours, are not revealed in this source.

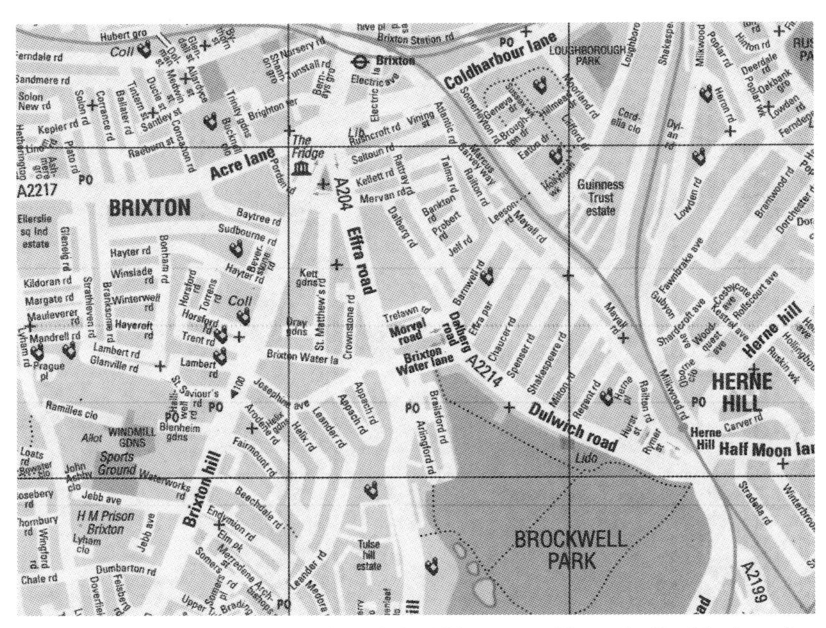

Brixton, from the *Nicholson London Streetfinder*. This map provides no details of the dynamics, character and tensions of the community. Indeed, Brixton is presented as a group of routes. Without roads there are no locations.

Another possible indicator is the distance between streets. The wider the gap the more likely that there are gardens of some size, although this is by no means an invariable rule. Thus, on the *A–Z Street Plan of Exeter* the close-packed streets of the Mount Radford area contain property considerably more expensive than those in St Thomas. The map emphasizes streets and thus ignores housing quality and character. Political canvassers and estate agents in Exeter are aware of detailed variations among streets, of a geography of zones and boundaries, of ownership and residence patterns that do not appear on any street plan.

Similarly, historical atlases that employ such maps generally fail to address the human dimension. *Mapping the Past – Wolverhampton 1577–1986* (published by Wolverhampton's Library and Information Services Division in 1993) included as its last map a section of the Ordnance Survey map for the city centre. This showed the ring road and the accompanying text referred to the clearance of 'sub-standard housing'. The accompanying destruction or dislocation of neighbourhoods left no trace in the map or text; the reader has to infer it.

The structure and density of activity of a city can also be ignored in typological maps that emphasize clarity, rather than scale and direction; for example the London Underground Journey Planner. This is based on a map devised in 1931 by Harry Beck, a draughtsman working for the

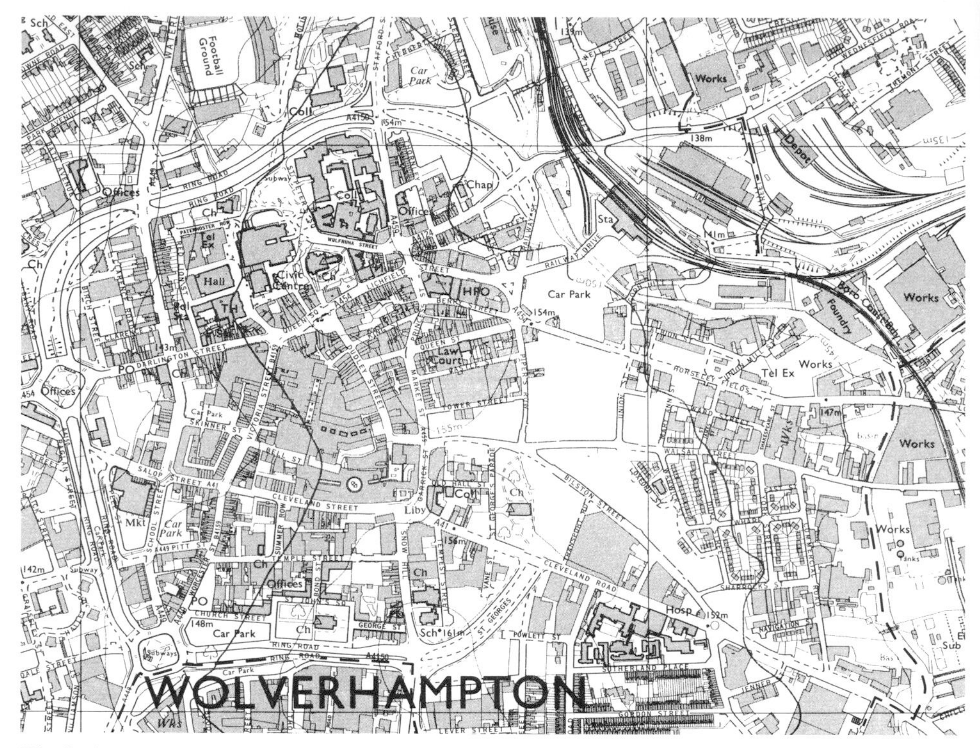

The Ordnance Survey map for the city centre of Wolverhampton. The new ring road is visible but not the damage and disruption it led to.

London Underground. Prior to that, the maps issued by the Underground railways were designed to be accurate in terms of distance and direction. The first such map, produced to show all the Underground lines, as opposed to those simply for an individual company, was issued in 1908 and depicted the lines superimposed upon a central London street map. Such a background was dropped by the 1920s, but the map still lacked the diagrammatic form that Beck was to introduce.

Beck's layout was inspired by scientific models, specifically by electrical-circuit diagrams, and depicted the lines as verticals, horizontals and 45° diagonals. His map was a success, and was used by London Transport for both station wall maps and pocket versions. Popular with users, the map offered an enlarged coverage of the central area, thus making it easier to understand routes and connections. The map also served another purpose. Some previous maps had had to exclude the outer, overground sections of the system or faced difficulties in including them, whereas the Beck layout included the entire system. By doing so, it shrank the apparent distance between suburbia and the inner city, and this achieved an important result at a time when suburban settlements, such as Edgware, were being developed. By making them appear closer, the Beck map ensured that movement there did not appear to be a case of leaving London. Instead, the ease of travel into the centre was emphasized, a visual effect that was encouraged by the use of straight lines on the map for the individual Underground lines.

The subject of a map reflects choice; so also do the scale, projection, orientation, symbolization, key, colour, title and caption. To imagine that there is a totally objective cartography is to deny the element and nature of choice and to neglect the assumptions present in choices, although these choices are often comprehensive within defined (and thus chosen) demarcations. Both assumptions and choices can require careful unpicking, as they entail subjective judgments, whereas the ideology of modern cartography, its *raison d'être*, is that of accuracy, which is generally seen as an aspect of objectivity; an impartial 'scientific' realization of reality. Most map-users see cartography as a science, a skilled, unproblematic exercise in precision, made increasingly accurate by modern technological advances. This approach is misleading, not least because it is based on a limited understanding of science. The limitations of the map-medium are more than 'technical' and non-controversial; the questions involved are more than merely a matter of which projection or scale to select, and with such choices seen as 'technical', rather than as involving wider issues.

It is possible, for example, to produce an accurate map of modern Italy that makes no allowance for the powerful fissiparous tendencies within the country. The administrative regions of Italy, regions that partake in the structure of state power, are mapped more frequently than the striking economic divisions between north and south, while Padania, the projected state of the separatist Northern League, proclaimed in 1996, appears on very few maps.

This essentially statist mapping was, and is, crucial, as the cartographic propagation of nations depends on a clear-cut identification of peoples and territory. To that end, space has to be understood as territory. The frequent reiteration of cartographic images of the state in, for example, rail and road maps and weather forecasting ensured that the shape and territorial outlines of states became clearly established. It is an educational process with a clear message about the natural way in which to order space. Such a process is enhanced by the comparable use of the shape of the state in advertising and in maps produced by commercial organizations.

In the 1980s and 1990s, the notion of objective cartography was challenged, in a series of arresting works. The most powerful of these was Denis Wood's *The Power of Maps* (New York, 1992). The chapter titles indicated the theme of the book: 'Maps Work by Serving Interests'; 'Maps Are Embedded in a History They Help Construct'; 'Every Map Shows

This ... But Not That'; 'The Interest the Map Serves is Masked'; 'The Interest is Embodied in the Map on Signs and Myths'; 'Each Sign Has a History'; 'The Interest the Map Serves Can Be Yours'. Wood's theme is that maps reflect and sustain power.

This treatment of maps brought together the iconic tradition of decoding paintings and other works of art familiar from art history, with postmodernist concerns about the nature of text and the contingent nature of authorial intention. Wood's book was dedicated to the memory of Brian Harley (1932–91), a British geographer whose example he acknowledged in the Preface.

Harley saw maps as essentially documents that contribute to the discourse of power, and that should be seen in that light. He treated cartography as a form of language, and maps as texts to be read and deconstructed in the post-structural and postmodern sense pioneered for literature, architecture and signs by Roland Barthes and Jacques Derrida.[2] Cartography was linked to, and to be understood with reference to, ontology, epistemology, iconography and reception theory. His concern with discerning the rhetoric of maps offered a new layer of meaning for the cartographic project,[3] and made it possible to relocate cartography in broader intellectual movements, and, in particular, in intellectual contexts other than those of the simple spread of knowledge.[4]

Moving away from the question of the accuracy of maps, Harley, especially in his later work, highlighted the nature of maps as instruments of power, in particular by drawing attention to their practical and symbolic role as assertions and communications of proprietorial and territorial rights.[5] His analysis of the connections between power, knowledge and the mapping process drew on the work of the influential radical French philosopher Michel Foucault, and, in particular, on the problematizing of knowledge and its relationship with power. Foucault sought to use the notions, symbols and language of cartography, specifically of space, boundaries and networks, in order to understand and make dynamic his views on the politics of knowledge. For Foucault, knowledge as struggle was to be understood in large part by reference to space: there were boundaries and spheres of contest; ideologies colonized terrains.

If power is about space, spaces were created through the exercise of power. Cartography could be seen as central to this process. For Harley, the map, far from being passive, and viewed, emerged as a form of control, even surveillance; its producer was a map-creator, not a compiler, who manipulated map-users and was effective insofar as he or she could do so. This manipulation rested in large part on a conformity to the cultural assumptions of the viewer, thus ensuring that the language of the map was more than that represented by its symbols.

This perception led Harley to focus on the morality of maps and the ethics of cartography, and to assert the need to fight back – to focus on the 'silences' in maps – the peoples ignored or marginalized – and to use maps to promote social justice.[6] In addition, an emphasis on the autonomy of map-creator and viewer led to a stress on their roles. These roles became a matter of scrutiny, designed to clarify the nature and impact of cartography. To use the language of the 1990s, map-users were to be empowered through knowledge.

Similar arguments have been advanced by some other recent writers on cartography, and they have acknowledged their debt to Harley. This was true both of some practising cartographers and of writers on the theory of maps. Harley was particularly concerned about the extent to which the cartography and mapping traditions of the imperialist powers had, in his eyes, distorted the historical, and thus present, cartographic treatment and understanding of those who had experienced imperialism, both outside Europe and within; for example the Irish. He claimed that the sense and naming of place of those who had suffered from imperialism had been appropriated, and that their understanding of territory and boundaries had been neglected.

Indeed, the mapping of colonies by the controlling powers was very much for their own purposes. When the French mapped Martinique and Guadeloupe after the Seven Years War (1756–63), their maps recorded the plantation system of a sugar, coffee and cotton colony and were also designed to provide information in the event of future hostilities with Britain. Mapping was carried out by engineers and was linked to a policy of fortification. On the maps, the names of owners were marked on plantations; not those of workers.[7] Thus, in Harley's view, the poor and the colonized had both been dispossessed by established Western cartography, and their cartographies had been neglected. The intellectual hierarchy that typecast non-Western cartography as primitive, or at least limited, was seen as serving a malign purpose. It was not so much that knowledge was power, as that power permitted a ranking of knowledges and that this ranking served the cause of power.

Similar concerns have been voiced by other cartographers concerned to recreate the past world of the non-Europeans or to present their modern world without European accretions. Jack Forbes, in his *Atlas of Native History* (Davis, 1981), argued that maps of the nineteenth-century USA were misleading in that they substituted the cartographic pretensions of the US government (and also the general perception of East Coast society) for the realities of native power. He sought to redress this 'mythological map', and presented his atlas as a cartographic catharsis:

this atlas represents . . . an intellectual process which we have to go through in the United States; a process of discovering truth free of ethnic bias and colonialist chauvinism.[8]

Forbes claimed that an objective cartography is possible, but that ethnic and cultural bias had to be countered to produce it; indeed, that the process of countering such bias was crucial to, and part of, accurate cartography.

The same ambition underlay post-independence mapping of Africa, past and present. However, unlike in the case of Forbes, the inevitable subjectivity of cartography was asserted in the best historical atlas of Africa, that by J.F. Ade Ajayi and Michael Crowder:

There is nothing fixed, or final, about a historical map. Maps are not simply a means of displaying historical evidence. They are interpretations of that evidence, often bringing into sharp focus the author's or editor's assumptions concerning the nature of the phenomena, process, or events under consideration.[9]

Similar themes were voiced by theorists, concerned to question the legitimacy of maps and map-making, and the role of maps in creating a misleading sense of natural territorial control:

When the fundamental importance of perceiving real and imagined space is compared to what passes for most mapping today, a huge separation is apparent. In our consumer society, mapping has become an activity primarily reserved for those in power, used to delineate the 'property' of nation states and multinational companies. The making of maps has become dominated by specialists . . . If you were entirely cynical, you could view the appropriation of mapping from common understanding as just another police action designed to assist the process of homogenizing 5,000 human cultures into one malleable and docile market.[10]

Cartography was thus to be reconceptualized, treated not as a science, with its often misleading claims for objectivity and progress, but, whether as a science or not, in multiple and changing intellectual and social contexts. According to this approach, cartography was to be seen as a discourse of power,[11] which related to an understanding of space itself as a prime cause and reflection of power relationships.[12] Countries became in part creations, essays in the structuring of space by power.[13] More generally, the modern conception of space has been seen as active in a double sense:

. . . they [maps] create social spaces while at the same time they are modes of spatial representation. They create those two aspects of spatiality through enabling two corresponding modes of connectivity. Maps connect heterogeneous and disparate entities, events, locations and phenomena, enabling us to see patterns that are not otherwise visible. They also connect the territory with the social order. Modern systematic maps rely on a standardised form of knowledge which establishes a prescribed set of possibilities for knowing, seeing and acting. They create a knowledge space within which certain kinds of understandings and of knowing subjects, material objects and their relations in space and time are authorized and legitimated.

This process is related to the rise of the modern European state:

The establishment of this new international space set in motion the process whereby the whole of the earth's territory could be mapped as one. All sites would be rendered equivalent, all localness would vanish in the homogenisation and geometrisation of space. In order to achieve the kind of universal and accurate knowledge that constitutes modern science and cartography, local knowledge, personnel and instrumentation have to be assembled on a national and international scale. This level of organisation is only possible when the state, science and cartography become integrated.[14]

A similar stress on power arose from attempts to understand the nature of map representation and use. An important recent work by Alan MacEachren, *How Maps Work* (New York, 1995), argued

that to more fully understand how maps work, we need to investigate mechanisms by which maps both represent and prompt representations. The communication paradigm took us a step in this direction but floundered due to a fundamental assumption that matched only a small proportion of mapping situations: maps as primarily a 'vehicle' for transfer of information. A representational perspective, in contrast, begins with an assumption that the process of representation results in knowledge that did not exist prior to that representation; thus mapping and map use are processes of knowledge construction rather than transfer. To more fully understand how maps work, then, we must consider the ways in which map-makers structure knowledge and the ways in which cognitive and social processes applied to the resulting cartographic representations restructure the knowledge and yield multiple representations.

The antithesis of communication versus representation does not exactly match the axes that structure discussion of objectivity and politicization, but there is a relationship. The idea of power embraces something of the authority of the gaze, the questions of who is looking, how and why. The power of looking is both enhanced and directed by the map, which in its apparent objectivity matches that of looking understood from the viewpoint of 'Cartesian perspectivalism', an approach that has recently been challenged.[15] Furthermore, the map is constitutive of a certain form of reality, not merely representative of it. The message and medium of the map is different from that of the written text, not least because the first offers a simultaneity of form and impression whereas the second provides a sequential organization and content.

The theory of knowledge construction offers only limited guidance to the processes and problems of map-creation, not least the exigencies and compromises that characterize the collection of data and the decision of what can be mapped. In politicizing these processes of choice and compromise, scholars such as Harley, in their search for conspiracy, simplified a complex situation, although they also enriched the discussion of cartography. They replaced 'can', as in what can be mapped and what can be seen in a map, by 'should', or argued that the question of what could and can be mapped was determined by cultural and political suppositions. In part, their very politicization of the theorization and analysis of cartography was itself a contingent political statement, an assertion of academic ideology. Harley, in particular, clearly distrusted the state, a body that he was inclined to reify and simplify. In selecting his target – the nexus of maps and governments – he, and others, sought to make coherent what was, and is, in fact, generally more diffuse.

The postmodernist interpretation of maps drew on a left-wing dislike and distrust of authority that neatly combined government (in its more malevolent formulation as the state), traditional map-making, conventional and established views on cartography, especially of its progressive and positivist character, and the notion of objective truth. Matthew Edney argued that 'the state continued (and continues today) to dominate map-making, both governmental and commercial, and to promote for its own reasons the empiricist illusion of cartographic mimesis ... Map-making was integral to the fiscal, political, and cultural hegemony of Europe's ruling elites'.[16] However, while it is true that the state does play a major role in map-making, it is less clear that governments actively 'promote' an interpretation of cartography. In addition, the notion that map-making was or is integral to hegemony requires careful analysis, and the very notion of hegemony is not always helpful to the understanding of an often more complex and diffuse situation.

Denis Wood saw the rise of map consciousness as linked to 'the continuous rationalization of management indispensable to the capitalist state', the latter clearly being understood as a negative. This was, at best, only partly true, for map consciousness reflected a range of developments, including increased travel.[17] Seeking to explain why reputable scholars, such as Harley, preferred the theatrical presentation of the Ordnance Survey in nineteenth-century Ireland as an instrument of political tyranny and cultural chauvinism, a view central to Brian Friel's successful play *Translations*, to his own scholarly demonstration of the more balanced institutional structure and practice of the Survey and of the degree to which English was not substituted for Irish in place names by a colonial state, John Andrews suggested that modern intellectuals were sympathetic to criticisms of government authority and that they liked the degree of attention the play focused on maps.[18]

The critical approach dissected by Andrews was scarcely novel in the 1980s and early 1990s. The search for structures and practices of authority and authorization behind apparent scientific impartiality, the deconstruction of science, was beloved of a generation of scholars. The process was not without considerable value, but it suffered from a number of weaknesses, including a tendency to state the obvious, a simplification of, and a failure to understand the nature of, power systems, and a preference for style over substance. *Épater les bourgeois* might be fun – it certainly gave life to lectures, or at least to lecturers – but it was, and is, strangely limiting. In addition, the recreation of non-European cartographies through the mediums of 'First World' research and publication was also a process not without its ironies, although it served as a reminder that modern 'First World' senses of time and space are both contingent and relatively recent in origin.[19]

Rather than searching for cartographic conspiracies, it was more valuable to underline the degree to which space was, and is, understood differently by contemporaries, and to see this as a central problem for contemporary map-makers and map-users, not as an opportunity for deceit. As Richard Dennis pointed out in his investigation of the social geography of nineteenth-century English industrial cities, ' "Insiders", such as slum dwellers, and "outsiders", such as medical officers, were unlikely to see things the same way, or even to agree on what they should be looking at'.[20] Thus discussion of maps and mapping is related to a general consideration of points of view.

If only one viewpoint were adopted for mapping, then this could be seen as a political statement or judgment, whether conscious or otherwise. However, the extent to which such an adoption might have arisen largely, or even entirely, from the availability, even survival, of the evidence has to be considered.

This is equally true of the present. Data availability and the lack of uniformity in data collection are not only historical problems, and this is also true of the way in which data are presented. For example, both past and present are segmented in a manner that reflects patterns of authority, and thus power. Political and administrative units are commonly used for the collection, presentation and analysis of non-political data; doing so at least defines the possibility of lack of uniformity in data collection. Thus data on literacy or health are commonly presented, as they are collected, by administrative unit, not grouped by, for example, regions of common economic indicators. Equally, most data that are not mapped are also organized in terms of political and administrative units. Furthermore, mapping in terms of different units, for example regions of common economic indicators, is possible, increasingly so. Provided all relevant information is spatially coordinated, data can be remapped without using political units, and such a process has been greatly eased by the digitization of data and the use of computers to process them.

Even if different non-political units, for example grid squares, can be employed for the base map on which data are to be presented, it is still the case, however, that all views are not equally heard. If the views of nineteenth-century slum dwellers are harder to establish than those of medical officers, other than, for example, through the writings of such officers, this is also the case today. Illiteracy is far lower than in the past, and a greater percentage of the population is listed in the records of the dreaded state, thanks to national systems of education, health and employment insurance. Yet that is not the same as providing evidence of opinion, while the records of the state classify in terms of criteria set by government. In addition, in some respects, the very multiplicity of opinions that characterize the modern world place a greater premium on the ability of some groups to present themselves prominently, and make it harder to distinguish the views of, not silent but quieter, less articulate or less-heeded majorities.

Paradoxically, this very problem of minority and majority opinions throws an interesting light on discussion about maps, because it is apparent that the notion of representational reality is widely held in at least 'Western countries'. Maps are taken to reflect reality and indeed are consulted daily on that basis, most obviously with reference to travel routes and weather forecasts. Issues of symbolization do not worry, and are not relevant to, the vast majority of map-users and are not explained to them. Thus an important aspect of the politics of cartography is that the very attempt to present maps and mapping as aggressively political in nature and intention is, itself, a minority opinion, whereby academics use the language of impartial enquiry to advance an agenda of their own, rather like the cartographers they sometimes criticize.

Harley's claim that 'Whether a map is produced under the banner of cartographic science ... or whether it is an overt propaganda exercise, it cannot escape involvement in the processes by which power is deployed',[21] can be amended by substituting 'analysed' or 'described' for 'deployed'. This is especially apparent because of recent attempts to use spatial distributions of, for example, wealth and poverty, or health and ill-health, in order to produce maps that are politically engaged, but without the nationalist bent with which propaganda maps are generally associated, and for which they are frequently condemned. The contents and tone of Ben Crow and Alan Thomas' *Third World Atlas* (Milton Keynes, 1983; 2nd edn 1994), for example, are alive to novel methods of presenting space and to the way in which particular emphases are created without necessarily any intention to do so. Thus, in this case, attention is drawn to the use of 'the Christian calendar' in measuring centuries.[22]

A 'Third World' perspective offers a way to analyse and present global space that is different from the conventional 'Western' approach, although in many cases 'Third World' perspectives are 'Western' creations, reflecting intellectual agenda, and publication and marketing strategies devised in the 'West'. Other strategies are present at more micro levels. In general, popular political movements have sought to create 'counter-space' in opposition to existing political structures,[23] and the same is true of efforts by radical or alternative lifestyle groups[24] to contest and redefine space. This produces a fluid situation in terms of what is expected from maps. The very notion of the community has been presented not only as a neutral topic for, or a means of, analysis, but also as a possible challenge to social and political practices and norms that can have a cartographic dimension. In addition, there is pressure for discovering and disseminating the perceptions of time and space held by aboriginal peoples,[25] and also for cartographic 'relevance', seen as both as an adjunct of 'justice', whether environmental, social, political, ethnic or gender, and as necessary if cartography is not to be sidelined. Thus Ferjan Ormeling observed, in 1995,

> Atlases nowadays have to compete with TV-soaps and computer adventures which pose crucial questions such as 'Will the hero be able to find true love,' 'Will he conquer evil,' or 'Will he solve the mystery?' Cartographic counterparts of these crucial questions would be: 'Does one control the environment in this region?' 'How far is this country from an ideal situation?' 'Do all inhabitants have equal access to the nation's resources?' or 'Do people here have better chances at success than people elsewhere, and, if so, at what price?'[26]

This approach uses the argument of 'relevance' to assert the case for an explicitly politicized cartography. The notion of competition with television, however, may be queried, and it can be argued that the crucial cartographic questions for most atlas-users remain those of relative location and transport routes. Nevertheless, these, and other, different ways of presenting space offer an appropriate way to consider maps and politics: a contro-verted rather than a conspiratorial approach, in which the multiple meanings of space are seen as challenges that ensure that no single cartographic strategy will be possible. Such pluralism is appropriate for modern democratizing cultures in which the notion of democracy has been expanded from a political issue centred on the franchise to the ideal or myth of the creation of a world in which the views and opinions of individuals are widely heeded and their expression not controlled by hierarchical power structures. Multiple meanings are also appropriate in an intellectual culture well aware of change and the contingent nature of analytical judgments.

This pluralism has been given added force by arguing that Harley and other deconstructionists make the same mistake as the cartographers they criticize by contrasting acceptable and unacceptable map-making, and underrating the extent to which there is, instead, a spectrum of standards.[27] This flaw is seen as similar to the professionalism of the traditional cartographers Harley criticized in discussing the scientific map,[28] and their hostility to what they see as inaccuracy and bias.

The dichotomous juxtaposition of allegedly acceptable and unacceptable map-making was also an aspect of Harley's argumentative approach, his search for the analytical 'other' that could bear the weight of his critique of traditional cartography. Yet there is a great danger in reifying this cartography. In both the content of maps and their presentation, traditional cartography was neither static nor uniform, nor unreflective. As so often, the quest for the new is valuable if it builds on, not denies, the old.

A critique of 'traditional' cartography is not the sole method and product of recent changes in the understanding of mapping. There has also been an extension of the sense of place, which, while not directly related to mapping, has consequences for the context within which it can be considered. Place has been reconceptualized to encompass also the analytic spaces in which knowledge is made both 'real', in a sense of specific locations, such as universities or laboratories, and conceptual. Thus knowledge is understood and disseminated in particular places, both material and metaphorical, with place as a common organizational term.[29]

The mappability, in conventional terms, of many of these places is limited. Given the understandable concern of cartographers with consistency and their unhappiness about lacunae in data or poorly defined subjects, symbols and data sets, it is scarcely surprising that this poses a considerable problem. Indeed, much of the new geography of place is best presented visually in diagrammatic terms: place is an aspect of model building rather than locational specificity, or, rather, if both pertains, the first is more important.

The recent re-presentation of cartography owes some of its intellectual energy to the related attempt to question and redefine the cartographic canon by drawing attention to non-European mapping traditions. This has also been an obvious and important field for critics of modern 'Western' cartography. Other mapping cultures, including earlier European understandings of cartography, are thrown into prominence both by the scholarly attention devoted to such traditions in the *History of Cartography* (Chicago, 1987–) project, especially those of East and South Asia, and by specific studies of indigenous notions of spatiality, for example those of Native Americans, Australian Aborigines,[30] New Zealand Maoris and Canadian Inuits.

The latter are presented as more holistic, more interactive with the individual and communal circumstances and needs of their production. Furthermore, they can be seen in a non-comparative context: one in which there is no sense of a single cartographic standard by which different cultures can be compared and judged. Indeed, this can lead to widely differing interpretations of what is a 'map'. It is not necessary to accept conventional Western definitions of maps, nor Western usage as a basis by which the products of other cultures can be diminished by being seen as 'maps'; failed realizations of the Western cartographic model. In short, politics is again involved in cartography, for it is through the projection (a nicely ambiguous and multifaceted term) of Western power that other cartographic traditions were ignored or diminished. In addition, only if different cartographic conceptions are comprehended can the spatialities of many other societies be fully understood.[31]

This, however, is not simply a matter of cultural relativism, of comparisons between European and non-European societies, because it is apparent that cartography has to take note of different accounts of spatiality and spatial links within particular cultures, societies and polities.[32] It is also important not to contrast European and non-European spatialities too readily. The role and location of sites of collective memory in non-European societies can be considered alongside symbolic spaces in European memory,[33] although it is true that the latter play only a limited role in European cartography. As ever, however, selection is present in mapping.

Non-Western cartography is not simply a matter of the spatial awareness of 'primitive' societies, 'primitive' being understood today not as a comment or stage on some 'great chain of being', Darwinian or otherwise, but rather as a society with different, and in some eyes more desirable, social, cultural and political arrangements and ethos from those in the West. Instead, it is also necessary to draw attention to cartographic traditions in bureaucratic societies with a developed literary culture other than those of Europe. This is particularly true of China, which in many senses was a leading seed-bed of cartography. Interest in mapping there developed early. The first known map in China dates from about 2100 BC and appeared on the outside of a *ding* (ancient cooking vessel). A map of a graveyard produced between 323–15 BC was uncovered in a tomb in 1977. Maps in China certainly became more common under the Western (or 'Former') Han dynasty (206 BC–AD 9), although very few have survived from before the twelfth century, when they were frequently used in various types of publication such as administrative works. By the first century AD the Chinese were employing both the scaling of distances and a rectangular grid system. The Chinese subsequently adopted the mariner's magnetic compass and the printing of maps before these were introduced into Europe.[34] The modes of transmission of cartographic ideas and techniques are difficult to assess;[35] certainly, the bold arrows that might be used to indicate influences on modern maps are inappropriate. However, it appears that Chinese advances such as printing by engraving on wood blocks were adopted by Islamic traders and thence passed to Europe. The Mediterranean world had witnessed significant advances in cartographic understanding, activity and techniques during the Classical period, but, thereafter, much of the knowledge was lost. Ptolemaic cartography was rediscovered in the same period as Chinese techniques spread.

Both the chronology and the pattern of developments and influences are obscure and have been differently interpreted. What is certainly clear, however, is that the Western cartographic project of the last half-millennium, with its Eurocentric assumptions and its relationship with the spread of European power, drew to a degree on Chinese roots. As with China, knowledge and interest combined to ensure that the resulting cartography had a particular focus; in the case of China, that of China and its immediate neighbours, the latter being understood and presented in relation to China.

Definitions of maps and understandings of cartography both involve issues of power. Politics stands as a metaphor for social processes that provide the context for cartography and mould much of its content and reception. Although some of the more strident claims about the role of power in cartography and of cartography in power can be queried, it is, nevertheless, the case that such issues should play an important role in discussion about the contents and purposes of maps.

2 Mapping the World and its Peoples

At the global level, the first and most obvious cause of contention about mapping is that of projection. This has to involve distortion: a projection is a flat (two-dimensional) representation of the globe and the (three-dimensional) curved globe is not flat. There can be no such thing as a 'correct shape' on a map projection, not least because maps have 'cuts', which occur along the edge of the map.

The most common representations of the world are rectangular. This reflects the nature of modern printing: the appropriateness of such images for the atlas page or double page and the extent to which single-sheet documents, whether maps or otherwise, are generally rectangular (as are computer screens). However, rectangular maps deprive the world of its circularity: they make each parallel and meridian appear as straight, instead of circular, and give the globe the misleading visual character of right-angle corners and clear edges. The very need to choose a projection emphasizes the degree to which choice is involved in the representational nature of maps.

A number of different projections have been produced over the centuries, to serve different purposes. The most influential, and the only ones to be adopted by 'developed', i.e. Europeanized, societies around the world, have been European. The world was first circumnavigated in the sixteenth century, and by Europeans. It is not surprising that many of the maps they then produced used a projection that made most sense in terms of the employment of the compass, and of maritime directions and links, especially in the mid-latitudes. Europeans needed to be able to sail great distances if they were to fulfil the commercial logic of distant possessions and trading opportunities.

In 1569, the Fleming Gerhardus Kramer, Latinized as 'Mercator' (1512–94), produced a projection that treated the world as a cylinder, so that the meridians were parallel rather than converging on the poles. The poles were expanded to the same circumference as the Equator, greatly magnifying temperate land masses at the expense of tropical ones. Taking into account the curvature of the Earth's surface, Mercator's projection

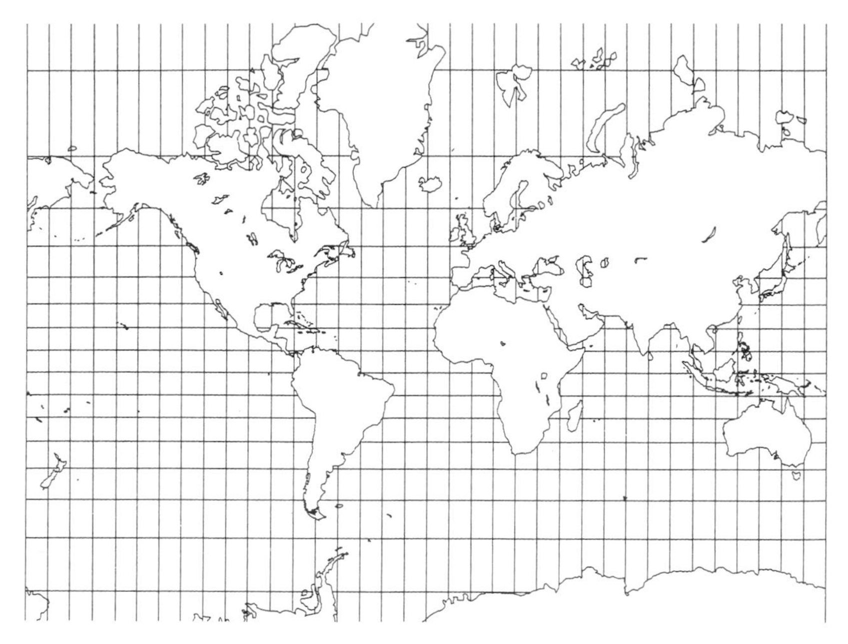

A world safe for sailors: a modern outline reconstruction of the 1569 map developed by Gerhardus Mercator.

kept angles and thus bearings accurate in every part of the map, so that a straight line of constant bearing could be charted across the plane surface of the map, a goal that was crucial for navigation. However, to do so, the scale was varied and thus size was distorted. Strictly speaking, Mercator is not a world projection: in its equatorial (as opposed to transverse or oblique) case, the poles are unshowable, because they would be infinitely large.

However, this was not a problem for European rulers and merchants keen to explore the possibilities provided by exploration and conquest in the middle latitudes to the West (America) and to the East (South Asia). The Mercator projection highlighted the imperial world of Portugal and Spain, and was an appropriate pre-figuring of the Spanish success under Philip II in creating the first global empire: the first empire on which the sun literally never set.

Unlike medieval Christian maps, the Mercator world was not centred on Jerusalem. Mercator placed Europe, which, to a European, both seemed most important and could be mapped most readily, at the top centre of his map, and gave the northern hemisphere primacy over the southern, both by treating the north as the top and by giving the southern less than half the map. However, a Mercator projection need not necessarily include more of the northern hemisphere than the southern. Similarly,

Mercator placed Europe at the top centre of his map, not of his projection: a Mercator projection might just as easily have the North Pacific in its top centre, with Europe split between left and right extremities; and it might just as well have the south at the top.

Mercator's is sometimes treated as the archetypal European projection. This is misleading, because, influential as it was and is, there was in fact some variety in the projections employed, although the notion of Europe as central and the northern hemisphere as on top was preserved. An equal-area projection was described by Johann Heinrich Lambert in 1772, and another in 1855 by the Scottish clergyman James Gall. Other early examples were those of Sanson-Flamsteed and Mollweide.

Yet concern with maritime routes and familiarity with projections of the Mercator type encouraged an essential conservatism in presentation, not least because one alternative, an equal-area projection, developed by J. Paul Goode in the 1920s, required cuts not only at the margins but also in the oceans,[1] and there was a cultural preference for flattening the globe onto a two-dimensional shape without any obvious joins or cuts except at the margins. However, in principle, a rectangular, circular or elliptical map is just as much cut as the kind of interrupted projection that is offered by a world map in two hemispheres, on which the cuts are very obvious.

The Van der Grinten projection, invented in 1898, continued the Mercator projection's practice of exaggerating the size of the temperate latitudes.[2] Thus Greenland, Alaska, Canada and the USSR appeared larger than they in fact were. This projection was used by the American National Geographic Society from 1922 until 1988; as such, it was very influential. The Society's maps were the staple of educational institutions, the basis of maps used by newspapers and television, and the acme of public cartography, for the period when the USA was the most powerful nation in the world, and, because earlier versions of maps enjoy a long life even when there has been a new and different edition, the influence of the Van der Grinten projection will long continue. In that projection a large USSR appeared menacing, a threat to the whole of Eurasia, and a dominant presence in the world that required containment. It was a cartographic image appropriate for the Cold War.

The geopolitical menace was abruptly reduced in the Robinson projection adopted by the National Geographic Society in 1988.[3] This offered a flatter, squatter world that was more accurate in terms of area. Thus, the Soviet Union moved from being 223 per cent larger than what the public was told it 'really is' – in the Van der Grinten projection – to being only 18 per cent more; Greenland from 554 per cent more to 60 per cent; Canada from 258 per cent to 21 per cent more; and the USA – further south than most British people appreciate – from 68 per cent more to 3 per cent less.

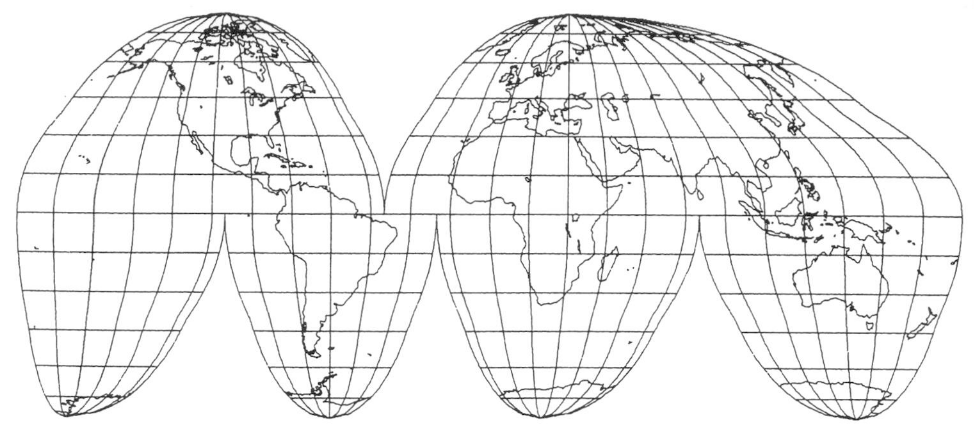

The Goode homolosine projection with interruptions used by Goode to preserve most landmasses uninterrupted, 1925. An equal-area projection that was influential in the USA, thanks to the support of the publishers, Rand McNally.

The Van der Grinten projection, invented in 1898. A projection of the world in a circle devised by Alphons J. Van der Grinten. The preservation of the general appearance of the Mercator projection, with a reduced area distortion, satisfied contemporary assumptions of the image of the world.

This last example can be seen as a piece of humility, as the product of a determination to lessen the extent to which the USSR outspaces the USA, or as the consequence of a Americo-centred cartography in which it was essential to create a projection that among the major states distorted the USA least. The variety of possible explanations, which can indeed be extended, reflects the difficulty of assessing causality in cartographic change. Arthur Robinson, a prominent academic geographer who had earlier been in charge of mapping at the American Office of Strategic Services during World War Two, had devised his projection in 1963, but it was not widely adopted until a quarter-century later.

The notion of map space as larger or smaller than a country really is reflects a widespread confusion. Of course, all maps are smaller than a country, and it is only in a work of fiction by Jorge Luis Borges that 'a map of the Empire that was of the same scale as the Empire' can be envisaged. If a projection is considered to be a transformation of a globe, then the 'real' scale (sometimes called the principal scale) of the resulting map is the scale of whichever globe you start with. Thus, to take the case of the Soviet Union, the real meaning in the last paragraph is 223 per cent larger than on a globe with the same principal scale. This, unfortunately, is cumbersome and to most readers incomprehensible, and it is easy to understand why such a form was, and is, not used. However, there was, and is, no excuse for another misleading change used to increase comprehension. Press discussion of the shift sometimes referred to 'Russia' when really it was the Soviet Union that was meant. The role of the National Geographical Society reflected an institutional dominance in the USA that was not matched in some other countries. Thus, in Britain, the Royal Geographical Society does not try to impose a particular projection on the British cartographic community.

It was scarcely surprising that American government spokesmen moved from employing maps based on the Van der Grinten projection to those using its Robinson counterpart, rather than the maps devised by the German Marxist Arno Peters. Using an equal-area projection similar to that of Gall, Peters devised his projection in 1967 and presented it to a press conference in Germany in 1973. It was in fact far weaker than many other equal-area maps, because it distorted shape far more seriously, greatly elongating the Tropics, so that, for example, the length, but not the width, of Africa was greatly exaggerated. Coastal shapes were thus considerably distorted and the standard cartographic images of continents, the iconic language of map shapes so important to map-readers, was changed. Distances on the Peters' projection could not be readily employed to plot data.

However, the reception of the Peters projection revealed the extent to

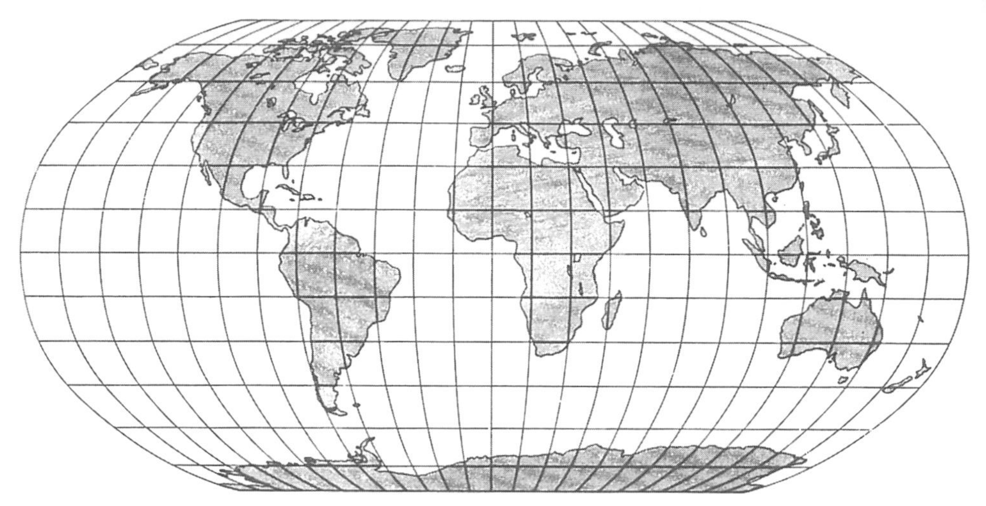

The Robinson projection, 1963. This was designed to offer the least possible area-scale distortion for major continents in an uninterrupted format map. It avoided exaggerated polar areas. Adopted by the US National Geographical Society in 1988, the projection then became commercially more successful.

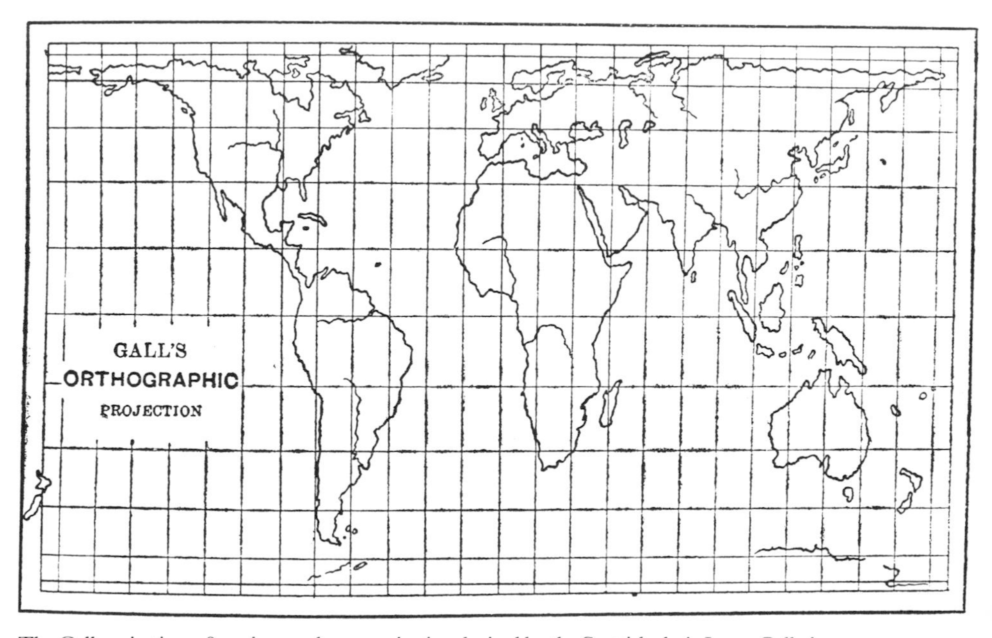

The Gall projection, 1855. An equal-area projection devised by the Scottish cleric James Gall. A cylindrical equal-area projection modified to obtain two standard parallels at 45° N and S.

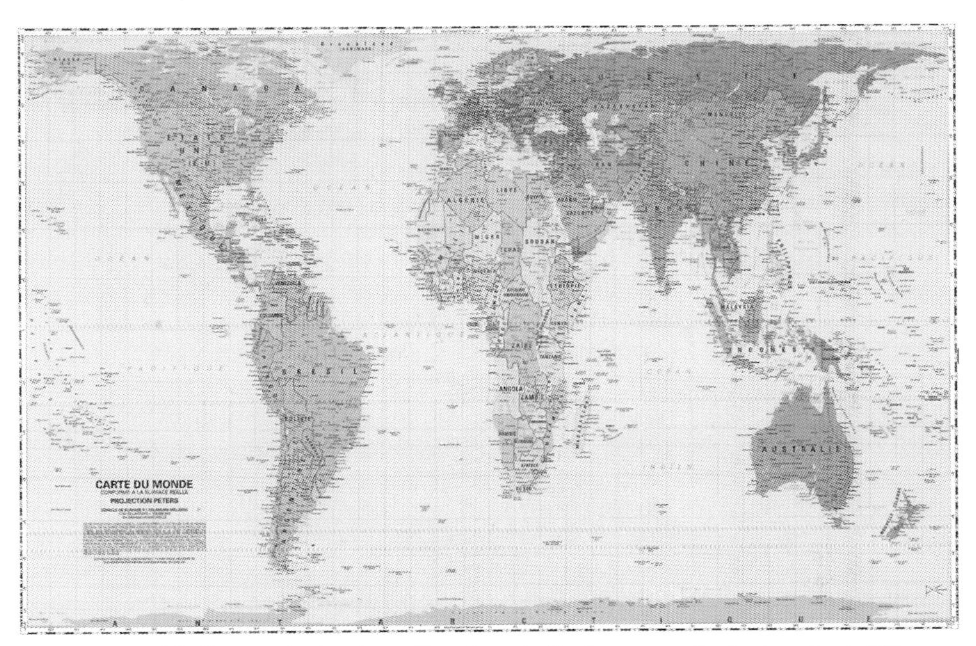

The Peters projection, devised in 1967, embodies the map as redistributive polemic. This projection was employed by Arno Peters as part of a distinct strategy for cartographic re-orientation, in which attention was redirected to regions that he felt had hitherto lacked adequate coverage.

which politics was more than simply a sub-text in projections and, indeed, more generally, in maps. Politically committed and an adept self-publicist, Peters portrayed the world of maps as a choice between his own projection – which he presented as accurate and egalitarian – and the traditional Mercator world view.[4] Arguing that the end of European colonialism and the advance of modern technology made a new cartography necessary and possible, Peters pressed for a clear, readily understood cartography that was not constrained by scientific cartography and European perceptions. With works such as *Die neue Kartographie/The New Cartography* (Klagenfurt and New York, 1983), published in New York by Friendship Press, Peters struck a chord with a receptive international audience that cared little about cartography, but sought maps to demonstrate the need for a new world order freed from Western conceptions. Peters' emphasis on the Tropics matched concern by and about the Third World, so that his projection was greatly praised by international aid organizations, particularly UNICEF (United Nations Children's Fund), international educational bodies, especially UNESCO (United Nations Educational, Scientific and Cultural Organization), and church bodies concerned with the Third World, such as the Papacy, Christian Aid, the World Council of Churches and the American National Council of Churches.[5] This was also financially beneficial: these organizations distributed over 60 million copies of the map and employed it to support their own points about the 'true' nature of the world.[6] The Peters world map was praised in, and used for the cover of *North–South: A Programme for Survival* (London, 1980): the 'Brandt Report' of the Independent Commission on International Development Issues, a prominent work, which conflated globalist perspectives, social concerns and redistributive strategies. Thus, a cartographic strategy for re-presenting the world was linked directly to the literature of concern and to political and ethical programmes for change.

Despite criticisms, the Peters projection continued as a politically correct icon. For example, in 1995 it was used for *The World Map* produced in an English version by Oxford Cartographers for a number of outlets including One Village, The World Shop, Oxford. The map carried the description:

> A map which represents countries accurately according to their surface areas . . . The map may look odd but it is more accurate than most. Maps produced in Europe usually show Europe and the north much larger than they really are – out of scale to the rest of the world. By contrast, this *Peters Projection* shows each country in correct proportion to each other. It is important to get this right, because maps have a profound influence on our understanding of the world around us . . . Although no conversion of a globe into a flat rectangular shape can be entirely accurate in every way, *Peters* is probably the most accurate and helpful map for today.

The text also included an attack on Mercator, set up as a straw-man alternative to Peters. The same was true of the undated version, also by Oxford Cartographers, sold in 1996 at The Exploratory Hands on Science Centre in Bristol. Its text closed with a section headed 'Fairness to all Peoples':

> By setting all countries in their true size and location, this map allows one its actual position in the world. In this complex and interdependent world in which nations now live, the peoples of the world deserve the most accurate possible portrayal of their world. The Peters Map is that map for our day.

Peters used his projection in a thematic global atlas he produced: *The Peters Atlas of the World* (Harlow and New York, 1989 and 1990). All the maps were at the same scale, ensuring that Africa, Asia and South America received more, and Europe and North America less, coverage than in traditional atlases. Thus all areas were to be equal, or, at least, treated equally.

The Peters projection, however, is open to serious criticism: it is far less novel than Peters claimed (he underplayed the similarities with Gall), while its shape distortion is serious, especially in the north–south dimension relative to the east–west, in the Tropics and near the Poles, its equal-area basis was challenged, and it was advanced with a dangerous dogmatism. Novelty, real or apparent, was not the only issue: the rectangular nature of the Peters map, its placing of north at the top and its use of a central meridian close to Greenwich ironically all proclaimed its conventional character.[7] A Pacific-centred Peters projection was, however, produced, first for a map for Malaysian Airlines published in 1993.

Other, less problematic, equal-area projections are available for atlases that wish to emphasize the Third World: Crow and Thomas' *Third World Atlas* (Milton Keynes, 1983) used the Eckert IV projection, which was also used by the *World Bank Atlas* (1980). Atlases have also been produced using projections that give a lower rate of distortion. The *Oxford-Hammond Atlas of the World* (Oxford, 1993) used an optimal-conformal projection for continental land masses that provided a 98 per cent accuracy rate.

It is apparent from the controversy over the Peters map that issues of cartographic choice over projections can be seen as politicized and can lead to political controversies.[8] This issue can be broadened. The question of 'which way up' the map should be arouses concern, especially in the southern hemisphere. In addition, there is the issue of what the map should be centred on, specifically that of Eurocentricity; ironically, the Peters world map located Europe in the middle of the map, even if Africa is at its centre. The relative role of land and sea also provides an issue for debate. In his *Atlas of the World with Geophysical Boundaries Showing Oceans, Continents and Tectonic Plates in their Entirety* (Philadelphia, 1991), Athelstan Spilhaus emphasized the sea more than other atlases. Spilhaus also challenged the control provided by edges and edging. He mapped what he termed 'a water planet': the world ocean uninterrupted by the edge of the map. To do so, Spilhaus produced a three-lobed map, centring respectively on the Atlantic, Pacific and Indian Oceans, with the map joined around Antarctica.[9]

The notion that Europe should be at the middle of the global map reflected both the role of Europeans in the development of cartography and the imperial power of European states, especially in the nineteenth century. Differently phrased, the notion reflected the dominance of European cartographic ideas, idioms and models. Europe's position was underlined by the international meeting of 1884, which chose the Greenwich meridian as the zero meridian for time-keeping and for the determination of longitude. This position was challenged, but essentially from within the European world. Thus several American cartographers in the nineteenth century constructed their maps with zero longitude passing through Washington or other American cities, and the notion of an American meridian was seen as a crucial aspect of national self-definition. In 1850, Congress decided that the Naval Observatory should be the offi-

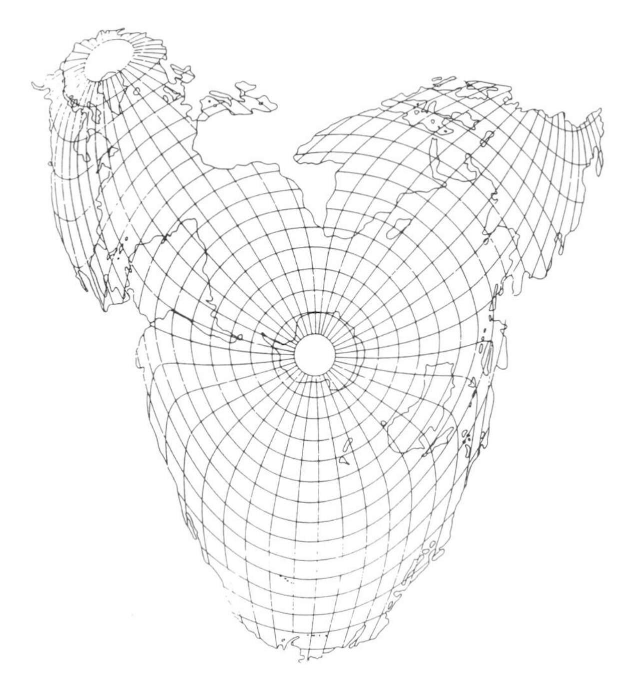

The Water Planet. 'Composite Shoreline Map', emphasizing the world's oceans. Athelstan Spilhaus developd a three-lobed map, with the lobes centred respectively on the Atlantic, Pacific and Indian Oceans. The first version was devised in 1942; the 1989 version was equal-area.

cial prime meridian for the USA, an act not repealed until 1912; and the French did not abandon the Paris meridian until 1911.[10] Even if the map were centred on the Greenwich meridian, understood as representing zero degrees of longitude, the USA would still be in a privileged position, because it is in the 'top left' quadrant of the map, which in Western culture is the customary place to begin scanning or reading a text, the point of origin.

The idea that Europe should be at the middle of the global map and the northern hemisphere at the top, with all the positive connotations implied by that double positioning, has been challenged by a number of maps. One of the most dramatic is *McArthur's Universal Corrective Map of the World*, distributed by Rex Publications (Artarom, New South Wales, 1979). Australia is in the middle of the map and the southern hemisphere on top. The tone of the text is combative, but its very extremism reveals it as likely to make little impact, and possibly this was appreciated by the compiler:

At last, the first move has been made – the first step in the long overdue crusade to elevate our glorious but neglected nation from the gloomy depths of anonymity in the world power struggle to its rightful position – towering over its Northern neighbours, reigning splendidly at the helm of the universe.

Never again to suffer the perpetual onslaught of 'downlander' jokes – implications from Northern nations that the height of a country's prestige is determined by its equivalent spatial location on a conventional map of the world.

This map, a subtle but definite first step, corrects the situation. No longer will the South wallow in a pit of insignificance, carrying the North on its shoulders for little or no recognition of her efforts. Finally South emerges on top. So spread the word. Spread the map! South is superior. South dominates!

Long live AUSTRALIA – RULER OF THE UNIVERSE!!

A less ostentatious challenge was that mounted by the growing practice in American atlases from the late nineteenth century of putting the western hemisphere at the centre, rather than on the left side, of the global map, although this location did not become common until the following century.[11]

To move from maps of the globe to those that are more detailed, it is readily apparent that atlases do not treat all areas of the world's land surface equally, and the same is true of the sea. It is most common for atlases of the world to depict the country, and indeed continent, of publication in greater detail than other countries and continents. As most atlases are published in Europe and North America this leads to those areas generally receiving greater attention. Furthermore, the 'customary' nature of world atlas contents, which critics might prefer to label conservative, ensures that standard assumptions about contents have developed. In spatial terms, these assumptions dictate an intensive treatment of certain regions, especially the USA and western Europe, more specifically the eastern states of the USA and northwestern Europe. By contrast, South America and Africa are treated in far less detail. Whatever the projection, such a coverage violates any notion of equal-area attention.

In addition, the standard pattern of world atlas contents bears little relation to the distribution of population. An atlas that reflected such a relation would devote considerable space to East and South Asia. In fact, such an emphasis cannot be found in atlases of the world. Relative to population, they greatly neglect China, India and, particularly, Indonesia, on the one hand, and exaggerate the importance of Australasia and Canada, on the other. Although to a less marked extent, the USA and western Europe also receive more attention than they merit, certainly in relation to Asia. As the distribution of space in an atlas is an implicit statement of relative importance, these emphases are relevant to the theme of this book, and to the impressions created on readers, but no attention is generally drawn to this issue within individual atlases themselves.

The *Peters Atlas of the World* (Harlow and New York, 1989–90) sought to counter the problem of spatial bias by ensuring that all its maps of parts of the globe appeared on the same scale. However the very use of the Peters projection, and in particular the visual attention devoted to the southern hemisphere, was problematic in demographic terms. Although the verticality of the Peters projection gave considerable prominence to India, while the emphasis on the Tropics as opposed to more temperate regions, gave China (which is mostly not in the Tropics but is further south than the USSR) greater coverage relative to the less heavily populated USSR than was usually the case, the prominence given Australia and the tropical regions of Africa and South America was inappropriate in demographic terms. The emphasis on the Tropics ensured that the southern states of the USA were more prominent than the northern ones, with Texas being especially prominent.

The issue of allocation of space can also be approached from other angles. Not only is a disproportionate amount of space in world atlases devoted to maps of Europe and North America, but these maps are also generally placed before maps of other continents. This very arrangement suggests a hierarchy of importance and a Eurocentric, or rather 'Western', primacy. Such a hierarchy is repeated within the 'West'. Wealthy areas are generally mapped more frequently and intensively than their poorer counterparts. Within Europe, far more attention is devoted in world atlases to the mapping of France, Germany, Britain and the Benelux countries, than to eastern Europe and the Iberian Peninsular. In many atlases one map will suffice for each of the latter, while France and Germany may be covered in greater detail. The scale of the coverage is an indication of relative importance, as is the presence of inset maps, for example, of London or Paris, but not of Madrid or Warsaw.

The *Reader's Digest Great World Atlas* (2nd edn, London, 1969) followed an initial spread giving a view of the Earth in the Solar System with a section on 'The Face of the World', providing maps made from sculptured models showing in relief how the world would appear to an observer at a point some hundred miles above the Earth's surface. The first spread, 'South Pole and North Pole', presented the latter from a point above Europe and also made the map of the North Polar world considerably larger than that of its South Polar counterpart. The following spreads were on Eurasia, Europe, the British Isles, Canada, the USA and Mexico, South America, Australasia, the Far East (including South Asia), northern Africa and southern Africa. Thus the British Isles received disproportionate attention, as did Canada, but the book was published in London, Cape Town, Sydney and Montréal.

The next section, the 'Countries of the World', began with the British Isles, followed by maps on Ireland, Scotland, southern England and Wales, northern England, the Arctic (a map in which the Greenwich meridian ran down from the North Pole in the centre of the map so that it was easy to read British place names and the Arctic was easy to understand in relation to Britain), Eurasia, Europe, the Low Countries, Switzerland, Scandinavia and the Baltic, central Europe, Iberia, France and northern Algeria (a reminder of colonial links), Italy, the Balkans, USSR, European Russia, the Volga basin, the Far East, east China, Japan, South-East Asia, South Asia, Punjab and Kashmir, plains of the Ganges, the Middle East and Afghanistan, the Levant and Jordan, southern Africa, central and east Africa, north and west Africa, Canada, Gulf of St Lawrence, the Great Lakes, USA, Middle Atlantic States, Pacific Coast, Mexico, the Caribbean, South America-North, South America-South, New Zealand, Australia, Indian Ocean, Atlantic Ocean, Pacific Ocean, the Antarctic.

The emphasis on western Europe was readily apparent, but the effect was more widespread than might be suggested simply by counting maps of Europe. For example, the map of the USSR, at one inch to 276 miles, offered only limited detail. This was supplemented by European Russia and the Volga Basin, both at one inch to 94 miles; but nothing comparable was offered for southern Siberia, Soviet Central Asia and the Caucasus, all important centres of population. New Zealand received a full page to itself at a scale of one inch to 79 miles, but the whole of South America was covered at one inch to 197 miles. Aside from the implicit statement that Chile, Brazil or Argentina were less important than New Zealand, the scale led to a crowding of place names in heavily settled parts of South America. In the page on southern Africa there were detailed inset maps for the Cape and for Witwatersrand, but not, for example, Katanga. Much of the Far East, certainly Indonesia, the Philippines and Korea, received little attention. They were all covered in the two-page map on the Far East, a map with a scale of one inch to 237 miles, and one that offered scant detail. The net effect, certainly if the map on the Far East was compared with those on North America, was to suggest that the Far East had a relatively low density of settlement.

These emphases were maintained in the next section, on 'The World As We Know It'. Thus, the two-page spread on 'The Ages of the Earth' was illustrated by one geological map, of the British Isles. The 'Evolution of Man' spread was supported by six maps, four of which were devoted to the origins of the peoples of the British Isles. The sole religious source-area to receive cartographic attention was 'the Bible Lands'. Exploration and discovery was largely presented from the perspective of Europeans, although Ibn Batuta's travels were also mapped. Nevertheless, the general effect was one of a world in which Europe and the Europeans were of greatest importance.

The cartographic emphasis on Europe and on wealth is not new. To take wealth, although a large number of American county maps were published in 1865–90, very few southern counties were covered because they were generally poor. A similar contrast was readily apparent in the large-scale insurance maps and atlases for about 12,000 American towns published from 1867 by the D.A. Sanborn National Insurance Bureau.

Poverty and the poor continue to receive a relatively small share of cartographic attention. Alongside the large number of modern maps produced by developers and estate agents detailing or depicting choice housing developments, their location and layout, it is unusual to find a work such as R.C. Prentice and G.B. Lewis' *Atlas of Housing Conditions in Welsh Districts* (Swansea, 1988). Its very format and appearance, and its publisher – the Housing Centre Trust, South Wales Branch – reflected a limited availability of funds and restricted marketing support. Golf courses are better mapped than poverty, although they are also of course easier to map, and it is more necessary to do so for the purposes of planning and play.

A 'Western' primacy is an accurate reflection of the world in one respect, in that it is a response to the distribution of per capita economic, and therefore atlas-purchasing, power. Maps and atlases are overwhelmingly purchased in the 'West' and the profit margins on their sales are also greater there. This can be readily seen with the pricing of individual atlases sold in different countries: titles sold in Africa, for example, have to be priced lower than in Britain. Thus the Eurocentricity, understood to include North America, of atlases in part reflects what sells or is believed to sell most copies. Atlases with different priorities frequently require subsidy; consequently, Spilhaus' *Atlas of the World With Geophysical Boundaries* (Philadelphia, 1991) was published by the American Philosophical Society. Yet such subsidies are most readily available in the 'West'.

Atlases reflect commercial pressures in other ways. The need for international sales, in order to recoup the high cost of production and to maximize profits, encourages publication in English, or another major international language: French or Spanish. Although atlases are translated for particular national markets, the language of first production or principal edition is of great importance as it conditions the place names used, and the space available for keys and captions. In addition, even though the titles and captions of maps, and the accompanying text, can all be translated, their contents are affected by the suppositions associated with the terms used in the original language of the atlas, as well as by the cultural suppositions of the people employing that language.

Nomenclature is an important issue, not least as formerly colonized areas seek to remove the linguistic impact of European imperial rule. This creates problems for cartographers. Readers outside India wish to find Bombay, Delhi and Madras, not Mumbai, Dilli and Chinai. Renaming is not simply directed against European imperial powers; it also seeks to reposition history, to make it truly past. For example, in 1996, the Hindu fundamentalist Shiv Sena government in Bombay/Mumbai wished to change the name of Aurangabad, a city named after the seventeenth-century Moslem Mughal Emperor Aurangzeb, and to call it Sambhaji Nagar, to honour a Hindu opponent of the Mughals. Thus, for a cartographer, renaming can be more directly political than the more general removal of European colonial accretions. It can relate to indigenous, or at least non-European, political projects, past or present, concerning which it can be difficult for a cartographer to show sufficient sensitivity and adequate judgment.

The 'West' is not only dominant in the commercial world of map-making, but also in that of state production, subsidy and distribution. The two are linked and, indeed, it is in part access to the opportunities of the commercial market that gives certain government cartographic agencies a particular edge. Such links are not new. From 1823, the extensive range of charts produced by the British Admiralty was offered publicly for sale in Britain. Mapping projects and institutions receive far greater public subsidies in the 'First' than the 'Third' World, such that both public and private sources of largesse and provision of opportunity are combined in one part of the world.

The modern role of satellite location-finding has given some government agencies an additional importance. The US Department of Defence developed a global positioning system (GPS) that depended on satellites. This was made available to civilians, but the accuracy of the signal was degraded by a process known as Selective Availability, so that positions thus obtained would be accurate to within 100 metres only 95 per cent of the time. Were this degrading not employed, the civilian signals would be generally accurate to within 15 metres. The technology is Western, and access to it requires an ability to afford, to understand and a wish to locate oneself with reference to 'Western' technology.

Criticism of Eurocentricity has come both from within and from outside the 'West'. Some Western map-makers have cast a critical eye on the suppositions underlying, and contents of, earlier atlases. A different cartography was offered from the 1920s by Marxists. This was true both of the maps produced in Communist states and of those produced by sympathizers elsewhere, for example John Horrabin in Britain. Marxist maps focused on issues of political and economic power in a manner that sought to draw out the implications of such power. For example, the economic links of colonial systems were presented in terms of imperialist exploitation. Similarly, in Marxist maps there was a focus on opposition to established power, both in colonial systems and in capitalist states; although their Communist counterparts were not exposed to such cartographic scrutiny.

More recent revisionism can be found across the range of atlas and map production, from works for children to academic atlases. Thus, Antony Mason, in *The Children's Atlas of Exploration* (London, 1993), responded to the controversy about Columbus excited by the cinquecentennial of his first voyage to the 'New World'. Mason discussed non-European explorers, and wrote of Columbus that he

found places about which he – and the rest of Europe – previously knew nothing. It was a discovery only from the European point of view. The fact that he went on to rename many of the islands, and to claim them for Spain, was typical of the behaviour of European explorers during the fifteenth and sixteenth centuries. Such arrogance and aggression has given the term 'discovery' a bad name.

Instead, Mason proposed a pluralistic and relativistic interpretation. He argued that 'the world is a jigsaw puzzle', a thesis illustrated in his atlas by a globe thus presented, and continued:

Each community in the world builds up its own knowledge about the world. For example, a European's understanding of the world is quite different from that of a Tuareg nomad in the Sahara Desert. We cannot say that one is better than the other. The Tuaregs' knowledge of the Sahara is far more useful to them than the European's idea of the world would be.[12]

This relativism was and is linked to a clear and open desire to redress past wrongs, to correct a cartographical inheritance seen as flawed.

In terms of maps, this emphasis is seen not only in a desire to increase the coverage of non-Western countries and peoples, but also to present them – in their present and their past – without concentrating on links with the West. This serves the purpose both of presenting a past and a present that are not primarily responsive to the impact of the West, and also of showing how they can be presented with minimal reference to the West.

At the academic level, this approach can be seen in a number of works. Thus Joseph Schwartzberg, the editor of the best atlas of South Asian history, wrote that

A final aim of this atlas is to contribute in modest measure to redressing a conspicuous imbalance in the presentation of South Asian history, which, despite the recent growth and vigour of the historical profession within the region, remains excessively preoccupied, in our judgment, with the impact of the West on South Asia and with the roles played by specific Westerners. As a simple illustration of a corrective, to be found nowhere else, we cite our map of 'Centers of South Asian Religious Movements Abroad'. Maps of Western missionary activity in India, by contrast, can be found in abundance.[13]

For atlases of the modern world, such an approach, for example, suggests the abandonment of the depiction of shipping routes, which were in many respects a European way of organizing and making sense of space: this was

The Globe as Jigsaw, from Antony Mason's *Children's Atlas of Exploration* (1993). Mason's presentation of the globe was designed to demonstrate the separate spatial understandings that co-existed in the world.

certainly the case with the routes that were depicted, routes that linked the world to Europe, and, more specifically, colonies to their mother countries. However, it is not plausible to reject the territorial legacy of colonialism completely, because, although colonialism only lasted for a comparatively brief period in some areas, for example Indo-China and much of Africa, the durability of colonial frontiers and political/administrative entities has been a notable feature of the post-independence world, and the political spaces thus created have continued to affect transport and economic activities. Furthermore, the ethnic impact of European colonialization was very important in the New World.

A growing concern about Eurocentricity has also been related to greater interest in indigenous peoples, their current interests and culture, and the way in which their cartographic traditions have been ignored. This interest has been particularly marked in areas of British and British-related colonial activity, especially the USA, Canada, Australia and New Zealand. In the first, a number of atlases devoted to particular Native American tribes appeared, for example J.M. Goodman's *The Navajo Atlas* (Norman, 1982) and J.J. Ferguson and E.R. Hart's *A Zuni Atlas* (Norman, 1985). Both were published by the University of Oklahoma Press and were evidence of the commercial possibilities of their subjects and, more generally, of American academic, popular and institutional interest in Native Americans. They presented a favourable view of their subjects, seen as essentially in accord with the environment and thus a ready contrast to the impression presented of European Americans; Native Americans were also generally presented as victims of European colonialization.

The two views were linked. If native peoples had a symbiotic relationship with the environment, they could not be seen as in some way having failed to develop its potential. The latter attitude had served to justify conquest and expropriation in the nineteenth century, an attitude given cartographic form by the association of European control with settled agriculture, while, in terms of land use, indigenous peoples were generally associated with what were presented as less intensive, and thus less valuable, practices. Expropriation was thus justified in order to permit exploitation: the latter was the necessary end of the human relationship with the Earth and one that was facilitated by the surveying of its riches. Maps thus assisted the utilization of resources; they were a natural corollary of the discovery of spaces and routes.

In the late twentieth century, there has been a major shift in consciousness. In Australia and Canada, indigenous peoples have recently been allocated a relatively large share in national atlases. Thus, *Australians: A Historical Atlas*, edited by J.R. Cram and J. McQuilton (Broadway, 1987), the nearest the Australians get to an authoritative national historical atlas, and the three-volume *Historical Atlas of Canada* (Toronto, 1987–93), both devoted more weight to indigenous peoples than earlier national atlases of those states, and did so in a more sympathetic fashion.

Greater emphasis on indigenous peoples was not simply a matter of devoting more space to them; there was also a major effort to understand their notions of space and mapping. This is true both of historic cartography, most obviously with the impressive and influential *History of Cartography* volumes edited by Brian Harley and David Woodward (Chicago, 1987–), and of discussions of current indigenous traditions. Both produce accounts of space and spatiality that are very different from modern Western notions, and, at times, these have been contrasted, throwing light on different spatial conceptions and cartographic realizations. Thus, in his *Le Carrefour javanais: Essai d'histoire globale* (Paris, 1990), Denys Lombard made both Dutch mapping of Java, their most important colony, and local views relative.

A focus on non-Western traditions of conceiving of and depicting space has generally been accompanied by a sense that 'Western' interpretations were and are, at best, one among many and, at worse, heavily flawed and compromised by the 'Western' intellectual and political tradition; more particularly, by the cult of science, a view of the environment as a sphere for control, and colonialism. To a certain extent, this approach is not very helpful, as it is not a case of comparing similar cartographies.

For example, the secularization of most knowledge occurred earlier in the 'West' than in other cultures. This can be seen as a weakness, inhibiting holistic understandings of life and the environment. Much, although by no means all, historic non-Western cartography was in part religious. It often focused on cosmology, as for example in India, or on attempts to make sense of the human environment in terms of the movements and activities of spirits, especially ancestors, as for example with the Aborigines of Australia, the Maoris of New Zealand and the native population in New Caledonia.

Such a sense of space was and is potent, not least because it includes time, offering a congruence of time and space, a four-dimensionality, giving strong meanings to landscapes. It is also arguable that indigenous senses of space were and remain potent because they have direct meaning for the people concerned. They are not abstracted from their experience by a process of cartographic professionalism, of specialized production and depiction; although it is unclear how far such indigenous senses of space can be presented as egalitarian. A romanticization of indigenous cartographic imagination may be inappropriate, as it neglects the social configuration of knowledge, access to it and power within such societies.

Westernization has certainly splintered the cartographic perception of non-Western peoples. In modern times there can be a difference between the representation of spatial views of the world (often religious and cosmological) and the view of geography that would be produced by non-Western peoples who have command of Western cartographic techniques. Westernization was not dependent on colonial control. Although Ethiopia was only briefly conquered by a European power, its maps reflect the imprint of Western patterns. The period of maximum Ethiopian political reaction against the West this century also saw the production of the *National Atlas of Ethiopia* (Addis Adaba, 1988), a work produced by the Ethiopian Mapping Authority, which was strident in its ideology, yet unsurprising in its cartography.

Aside from indigenous peoples being seen as relicts, resisters and victims of colonialism, there is the more general problem of the mapping of ethnicity. This is a politically charged issue, not simply because of issues of ethnicity within states, but also because the notion of the nation-state, as the most legitimate form of political organization, has necessarily directed attention to ethnic boundaries and their mapping.

Race, a categorization based on physical characteristics, is notoriously difficult to define; religion and language are more straightforward, although they still pose important problems and are politically contentious. Ethnicity draws on all three categories, not least because of the role of image and self-image in definition. Thus an ethnic group generally has a common language or religious tradition, and these are important to its shared culture and traditions and to their distinctive nature. Ethnicity is no longer a factor in nationality in most countries, although there are important exceptions, not least in 'liberal' states, as Turks in Germany are aware, while political culture and practice may be such that enjoyment of the full benefits of nationality is indeed related to ethnic considerations.

Race, ethnicity, religion and language are frequently discussed and applied as if they were precise. This was, and is, far from the case. All of them are problematic as systems of classification and thus, even if relevant data can be established and plotted precisely, mapping is far from simple and involves choices that will be contentious and can be seen as political. 'Breed' as a categorization has even more unfortunate connotations. To take race, which is an attempt to apply to humans a taxonomic classification below the level of species, any definition encounters problems: no race possesses a discrete package of genetic characteristics; there are more genetic variations within than between races; and the genes responsible for morphological features, such as skin colour, are atypical.[14]

These scientific issues interact with institutional practices. The systems of classification used for race vary, not least in the treatment of bi-racial marriages and of mixed-race people. The practice of such marriages and unions helps to underline the very fluidity of the situation: unless entrenched through endogamy, demography undermines racial classification and thus racial mapping. As a consequence, whichever concepts are employed in the classification and mapping of race, there are major problems with the consistency of the data. Races are constructed as much as described, and mapping plays a role in such construction.

Given that the definitions of race and ethnicity are fluid and contested, it is scarcely surprising that mapping them is problematic. This is especially so because a frequent practice in such mapping is to produce homogeneous blocs of colour separated by clear boundaries. In short, in many maps, race, like language and religion, is treated as a creator of clear-cut units that are as readily apparent in maps as states. An equivalence between the two is asserted by means of using a similar cartographic language. Such a process then encourages a sense on the part of the average map-user, if such a term can be used, that the boundaries ought to correspond, and that, if they do not, there is a clear anomaly.

This approach was used by nationalist polemicists, for example protagonists for the creation of nation-states or for territorial gains by existing states, in Europe in the late nineteenth and early twentieth centuries. It is

The first London Underground Railways folding pocket map, issued free in 1908. An accurate reproduction in which lines were related to a central London map.

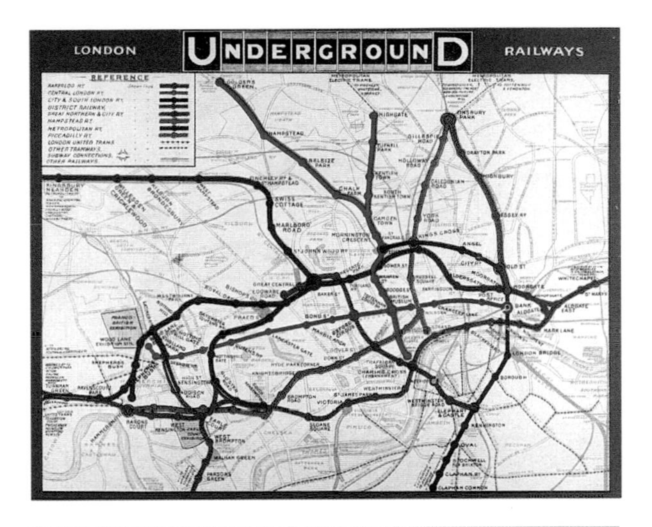

The 1927 London Underground map designed by F. H. Stingemore. This map recorded the expansion of the network into the suburbs.

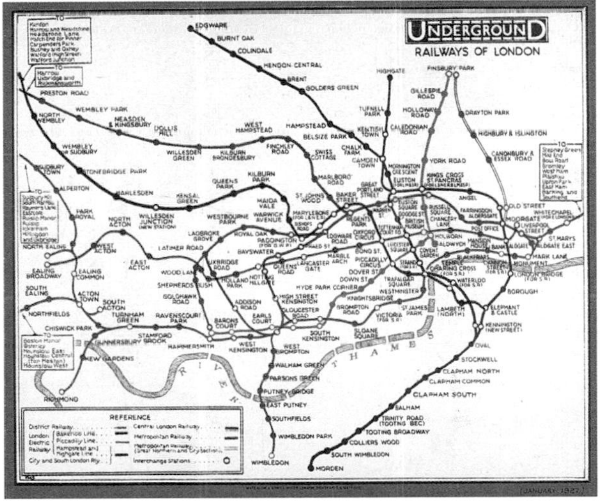

The 1933 Underground map by Harry Beck. A diagrammatic approach that offered clarity rather than a close relationship to actual directions and locations.

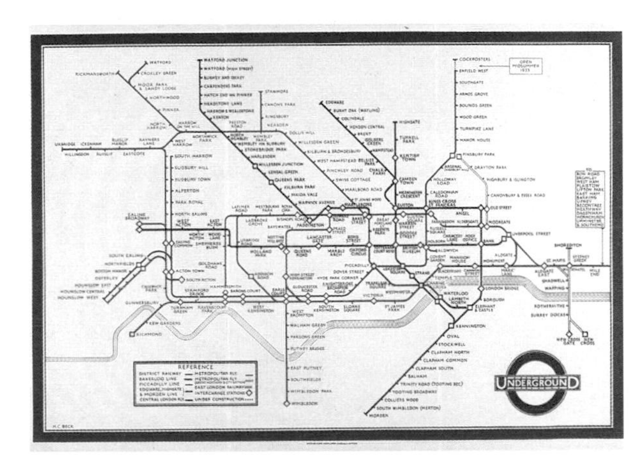

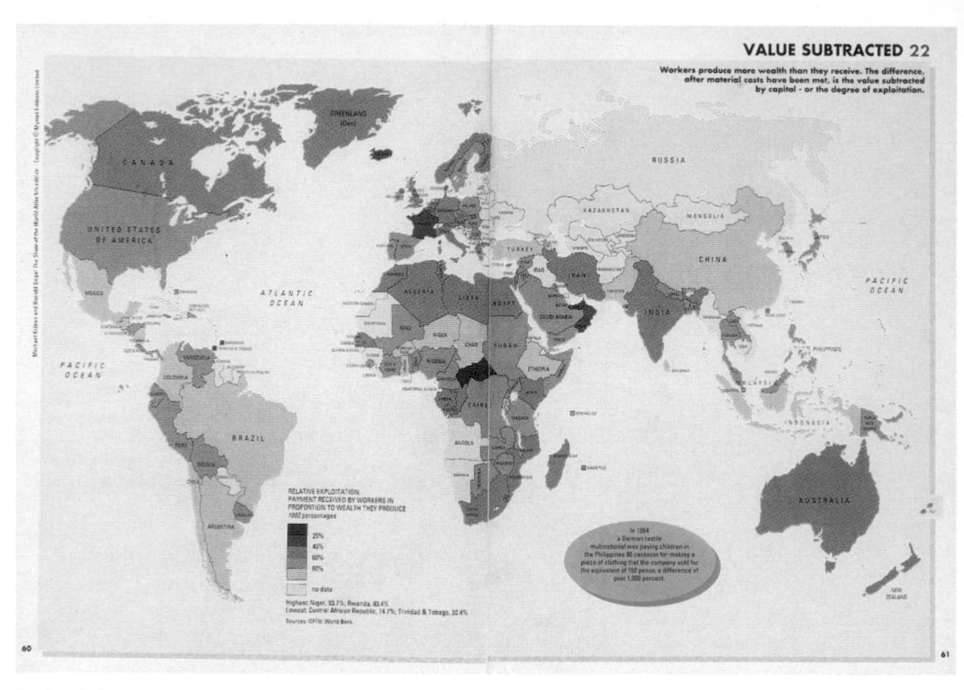

'Value Subtracted'. A tendentious mapping of relative exploitation from Michael Kidron and Ronald Segal's *State of the World Atlas* (1995). A crude but vigorous map in an atlas characterized by hostility to capitalism.

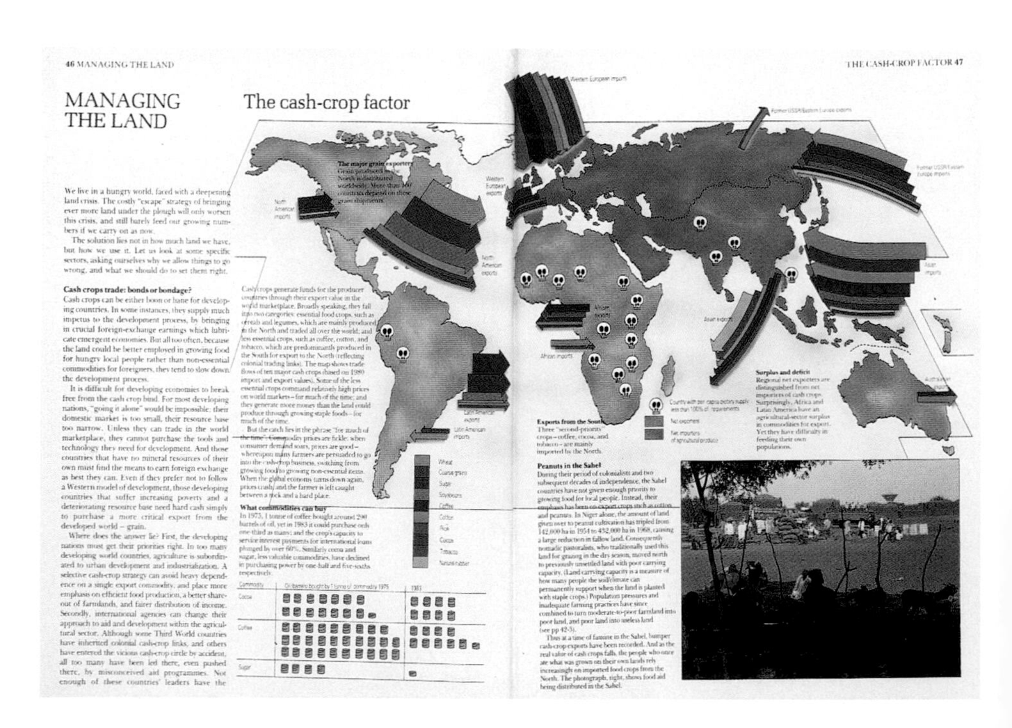

Opposite below 'The Cash-Crop Factor' from *The Gaia Atlas of Planet Management*. A presentation of the global economy that emphasizes the problems arising from the export of agricultural produce. The use of death's heads dramatizes the issue, but does not add to the analysis.

Right 'Change in Median Household Income, 1980–1990', from John Mollenkopf's *New York City in the 1980s: A Social, Economic and Political Atlas* (1993). A very different account of the city than that provided by a road map.

Below right Changes in 'unearned' income, 1980–1990, from Mollenkopf's *New York City in the 1980s*. The financial benefits of the decade were concentrated in white upper-middle-class areas. Dividends, interest and rent as a spatial indicator of class society.

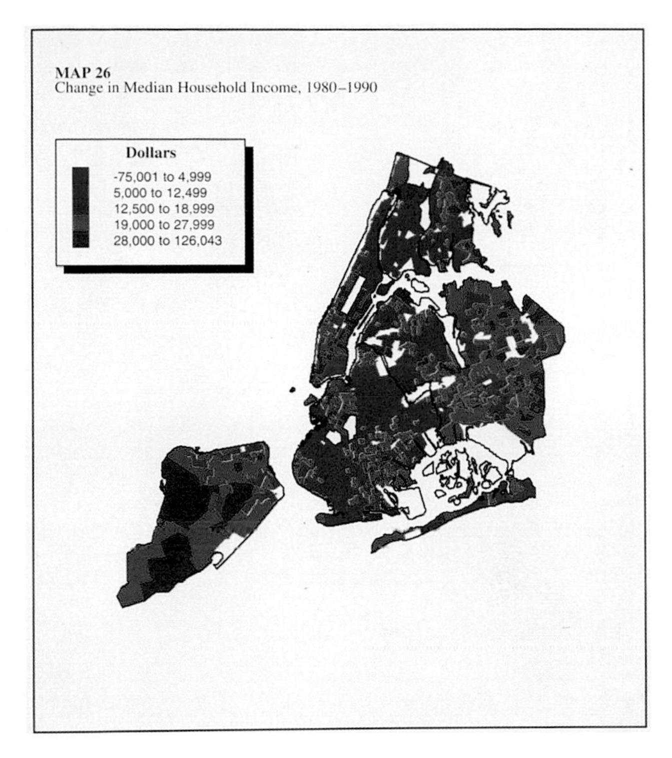

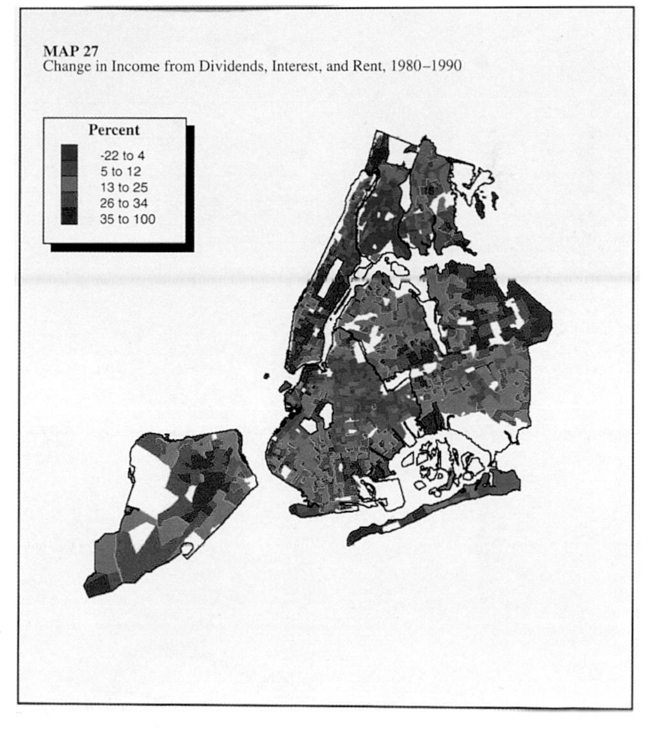

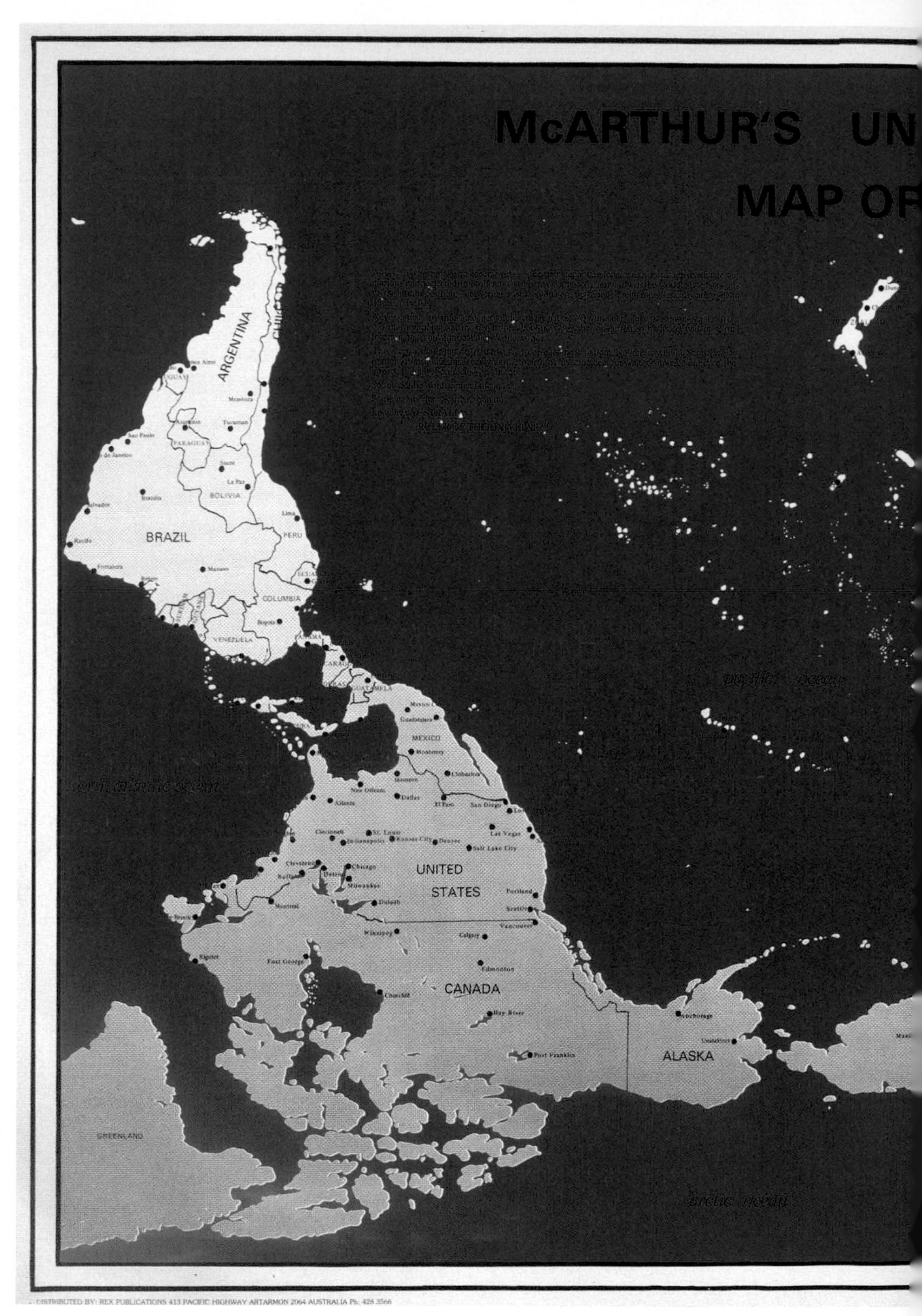

McArthur's Universal Corrective Map of the World (1979), one of several that puts the southern hemisphere on top, offering powerful challenges to conventional assumptions.

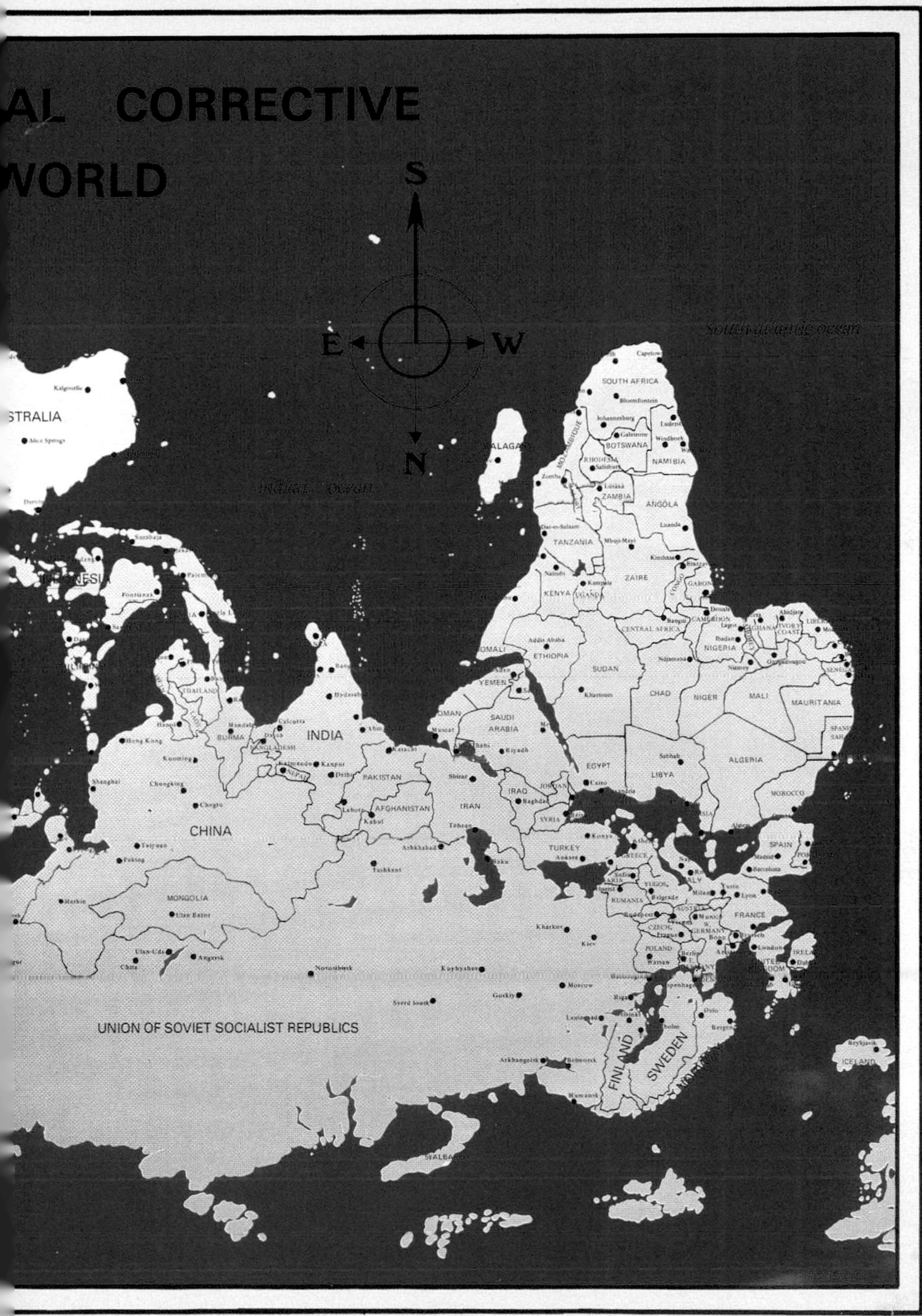

Created by S. McARTHUR 4 Alana Lane Frankston 3199 (03) 787 6605

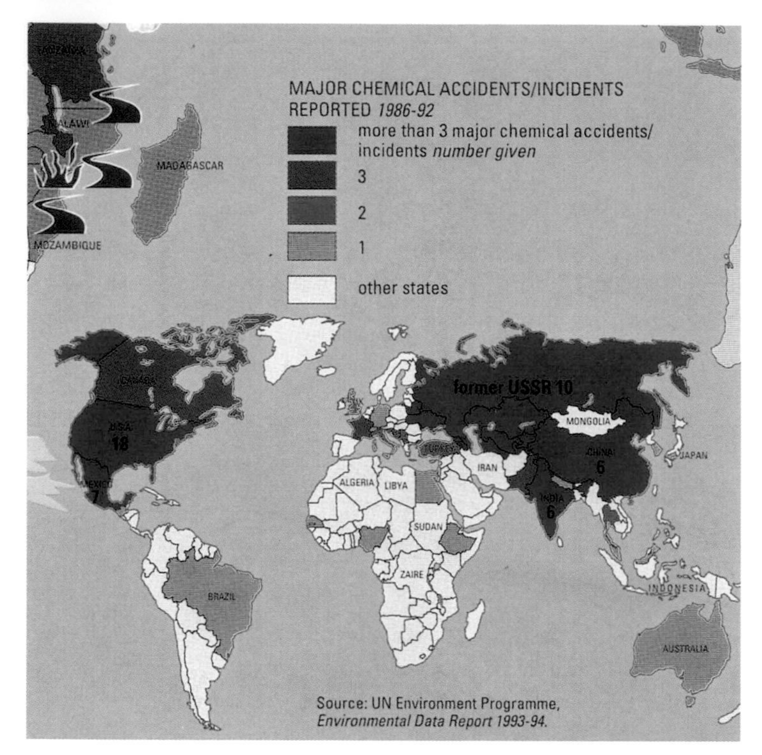

'Major Chemical Incidents/Accidents' from Kidron and Segal's *State of the World Atlas*. The use of red is intended to dramatize the extent of the problem, but 'major' is a subjective term and there is no indication of the relationship between accidents and total activity.

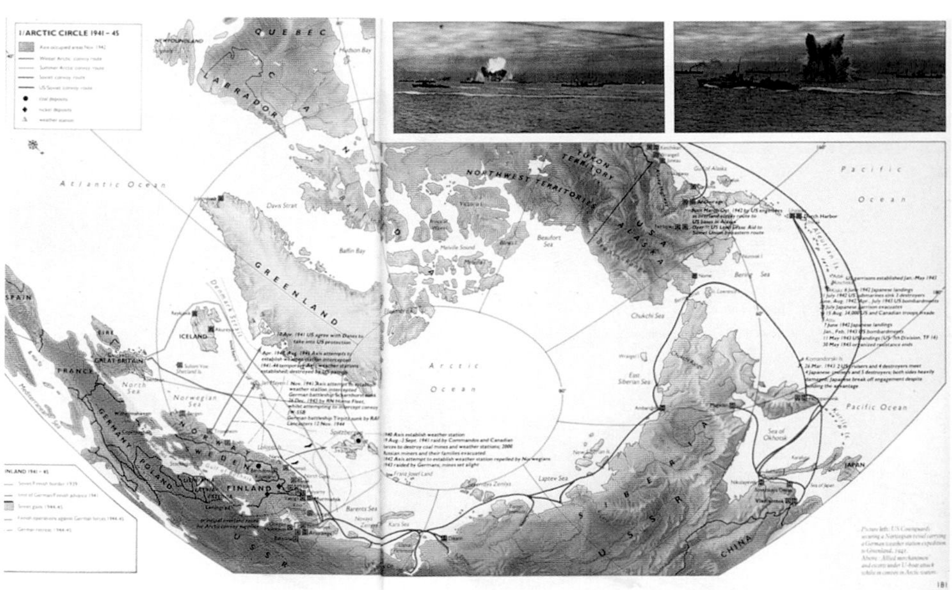

'War in the Arctic Circle' from John Keegan (ed.), *The Times Atlas of the Second World War* (1989) directs attention to a region that is generally ignored in the mapping of the war and adopts a projection that links conflicts in the northern Pacific and off Norway. It fails to point out however that Japan and the USSR were not at war.

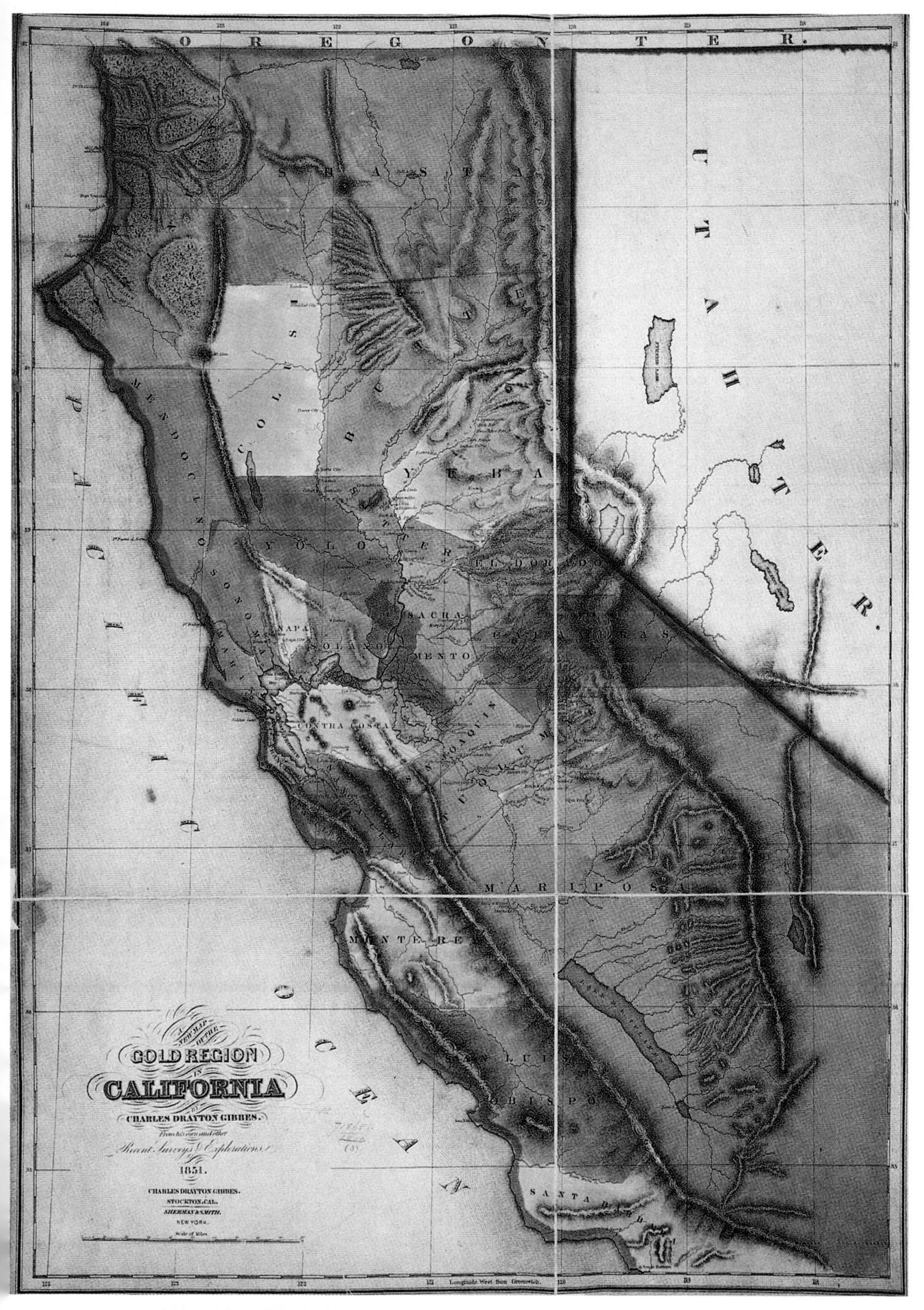

A New Map of the Gold Region in California by Charles Drayton Gibbs (1851) was an example of the use of the map in order to assist exploitation. The native population was ignored.

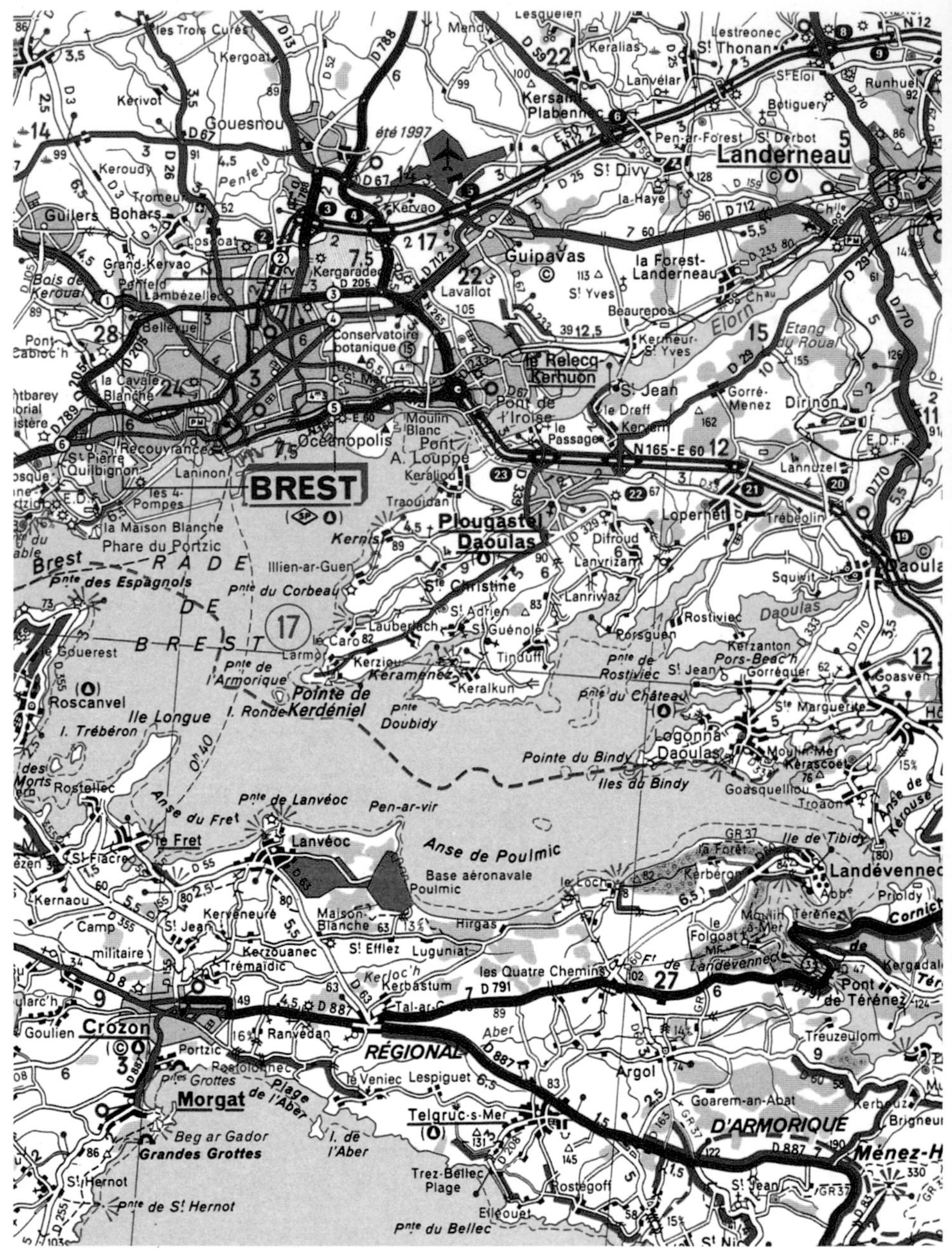

Brittany, from the Michelin Carte Routière and Touristique series. A region known best for its coastline presented with an emphasis on the major road routes. Colour provides a hierarchy, emphasizing major roads, while railways are less prominent. The Michelin series integrates France: different environments are made familiar by the employment of the same cartographic conventions and symbols, and there is no allowance for Breton or Corsican nationalism.

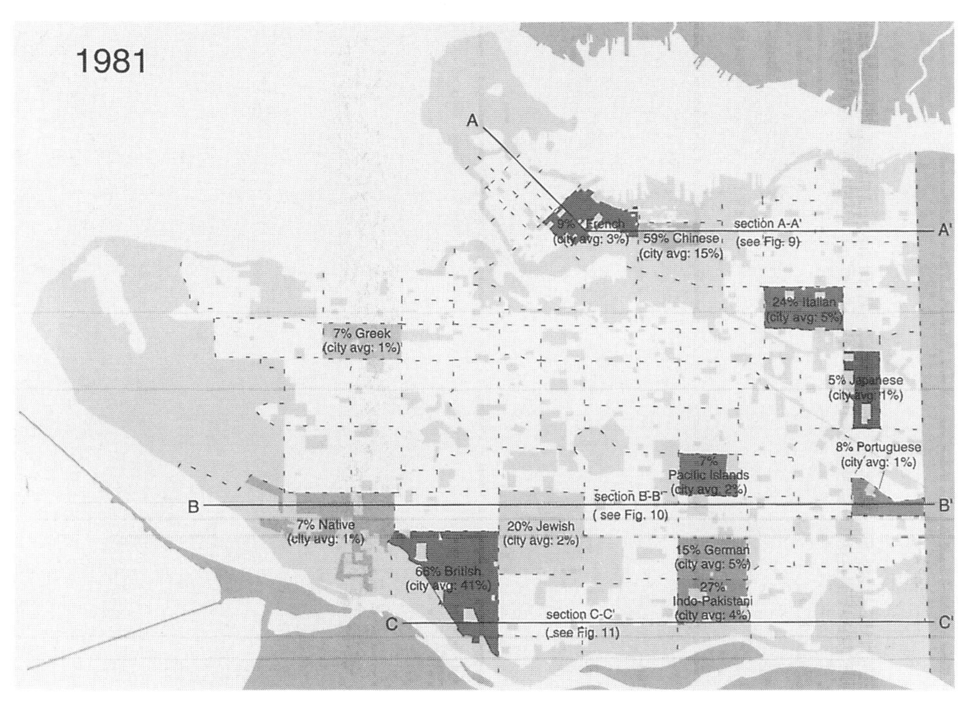

Mapping Ethnicity. Largest ethnic origin group by census tract, 1981, from Bruce Macdonald's *Vancouver. A Visual History* (1992). A mapping of ethnic concentration that indicated where each major ethnic group was most concentrated, but without suggesting that this led to homogeneous blocs.

also used today in discussion, for example, of the position of the Kurds. Furthermore, the 'naturalness' of ethnic, racial, religious and linguistic blocs appear to correspond to the 'naturalness' of political divisions. Each supports the other in creating exclusive spatial notions of identity. Keys to maps generally make no allowance for anything other than majority groups, and this is true whether race, language or religion is being depicted. Aside from the error this presents, it also creates a misleading sense that such a homogeneity is the norm, and that any areas that have to be presented as heterogeneous are atypical.

Homogeneity is generally implied by downplaying or ignoring minority communities, but in some cases it entails doing so to the majority community. This was particularly the case during the apartheid regime in South Africa.[15] Numerous substantial Black settlements, particularly townships and squatter settlements, were ignored or minimized on maps of South Africa. In part, this reflected the difficulties of mapping such settlements, especially squatter settlements, but this was not the major issue. As with differences in the accuracy of census data between ethnic groups in South Africa, the fundamental issue in the mapping was a concentration on the White population. Thus, in maps, small White-dominated towns, especially in rural areas, were given a misleading prominence.

Mapping techniques, such as unit-dot maps or the employment of interdigitated diagonal bands of different colours, exist to display heterogeneity, and to create an impression of a nuanced reality, a spatiality of emphasis not consistency, but they are generally not used. One possibility is the use of highest ethnic concentrations by census tract, as in *Vancouver: A Visual History* (Vancouver, 1992). This mapped the single highest density of each ethnic group that made up more than 5 per cent of a census tract's population. The percentages given in brackets, showing the percentage of each group over the whole city, served to indicate the degree of concentration or assimilation of each group shown on the map when they are compared with the percentages of the densest area. New groups, visible minority groups, and groups retaining the culture of their home country, for example the Chinese, show a tendency to congregate, while for ethnic groups that have assimilated, such as the Germans, the data has less meaning. The value of such mapping is that it does not suggest the existence of homogeneous blocs. Unit-dot maps, where one dot represents a fixed quantity and demographic diversity is shown by having different coloured dots on the same map, are very confusing, but this very confusion is instructive; it is an accurate reflection of the reality of racial, ethnic, religious or linguistic diversity. The use, instead, of homogeneous blocs is misleading. Given the role of religion, race and language in reflecting and fostering division, the dominant current pattern of mapping these subjects is unfortunate.

3 Socio-Economic Issues and Cartography

The physical world

Any apparently homogeneous territorial space is problematized if attention is devoted to its internal differences. Traditionally, a major divide in the mapping of the globe, continents, countries or smaller areas, as seen particularly in atlases, has been between maps of political or administrative boundaries and maps concentrating on the physical geography of the area in question. The latter, of course, emphasize difference, particularly of height, but the differences appear 'natural', not controversial.

In fact, there are issues of choice involved in presenting physical geography, and the controversies about these choices can be politicized. A choice of contour intervals and shading and an emphasis on peaks that present a country as mountainous makes a different statement about it than one that minimizes the impression. Similarly, there is no clear standard by which to establish the number of rivers that should be shown on a map or the quantity of marshland. Again, to emphasize either, for example by mapping all rivers above a certain length, width or average flow, or by mapping tributaries, and often minor ones, alongside mighty rivers, creates an impression that triggers the connotations associated with riverine or marshy character.

Thus the *Times Atlas of the World* (London, 1968) was assiduous in indicating marshland; indeed, the key included separate symbols for 'saline marsh', 'marsh, swamp', 'swamp, flood-area', and 'mangrove swamp'. As a consequence, areas such as the Guadalquivir Valley below Seville, the Costa Blanca south of Alicante or the Atlantic coast of Iberia between Faro and Huelva appear very different from their depiction in atlases of the period that ignore or minimize marshes.

It is also difficult to assign meaning to 'physical geography' where there has been human interference, as in the damming of rivers, the conversion of lakes to reservoirs, reclamation from the sea, and the drainage of marshlands. Nevertheless, physical geography is commonly seen as akin to organic. Thus, for example, maps and supporting text can emphasize that surface features reflect geology. This contributes to a situation in which there also appears to be a flow of influence from physical to nonphysical maps, and, indeed, in which the influence can be presented or seen as deterministic. Such a tendency was especially strong in atlases in the early decades of the twentieth century. At that period, an intellectual belief in strong environmental influence combined with cartographic techniques that made it easier to show terrain clearly and comprehensively than it had been in the nineteenth century.

Physical geography has since come to play a smaller role in maps and atlases, certainly in those of the world in the second half of the twentieth century, than hitherto. That reflects a lesser interest in environmental influence, which is a consequence of intellectual and economic shifts, including a decline in the role of agriculture, and a sense in the 1950s and 1960s that humanity could mould the human environment. Roads rather than rivers or mountains came increasingly to locate routes and boundaries, both in the mind and in maps. In townscapes, rivers became less important, as they were increasingly bridged or diverted to underground channels.

Nevertheless, physical features continue to play a role in maps, both specific maps of physical geography and, although generally to a lesser extent than hitherto, in other maps. The positioning of physical maps, usually before those dealing with other topics, can also suggest a degree of environmental determinism, and might imply the naturalness of what is more generally mapped and of what follows in the atlas, for example political boundaries. Such a positioning is important because the crucial character of an atlas, one that distinguishes it from a collection or selection of maps, rests in this arrangement of its contents in a logical fashion, and this very arrangement enhances the value of the maps and makes sense of them. James Akerman has argued that this logic has to be identifiable in order to supply meaning to the structure of an atlas, to make it work as a narrative form of cartography.[1] It is, however, far from clear that most readers approach an atlas seeking a narrative, but, if they do, it is provided in many atlases by placing the physical maps first. At the most basic of levels, the atlas thus offers a two-stage account of the evolution of the humans' Earth: first to the current physical state, and then to the human landscape.

Physical geography is generally seen as objective and maps that emphasize physical features are accordingly regarded as non-political. This is far less the case with maps that focus on social and economic features, although, again, they are generally not designed to appear controversial. Nor are most maps of social and economic features designed to support an argument; they are intended to be descriptive, not didactic. Such features are receiving greater attention, proportionately, in modern maps and atlases as the space and prominence devoted to physical geography declines.

The use of maps of social and economic features or characteristics, in order to make didactic points, however, is common. Furthermore, there is a cartography on the subject that sees the spatial distribution of economic characteristics as crucial causes, aspects, and products of a system of power. Economic and political are fused in this analysis, the workings of the market seen not as an impartial deistic presence, but as an adjunct of power relationships that are generally presented as far from benign; indeed, with their very definition as power relationships carrying such an implication.

This cartographic politicization of economic spaces is best presented in the two-volume *Atlas mondial des multinationales* (Paris, 1990–1), published by Reclus in the series 'Collection dynamiques du territoire'. The first volume, *L'Espace des multinationales,* by Pierre Grou, shows how economic space and, with it, polarization emerge. The maps are accompanied by a frequent feature of French cartography, '*scémas*', figures that indicate the principal elements of the relationship and their character, the latter generally achieved by means of the ubiquitous arrow, which is frequently given prominence by size, colour and dramatic usage. This device serves to visualize, emphasize and explain the power relationship accompanying the distribution depicted in the map. However, by making space an adjunct of the relationship, such figures offer a crude, and often misleading, unidimensional cartography: the use of arrows and the emphasis on causality can diminish an emphasis on space and spatial characteristics.

The second volume, *Stratégies des multinationales* by Claude Dupuy, Christian Milelli and Julien Savary, provides maps of the activities of leading international companies, such as Sony and Matsushita, and of sectors that reflect spatial choices, for example Japanese research institutions in Europe and French factories in Brazil, and seeks to explain the spatial understanding and strategies of leading companies, with sections and maps, for example, on 'L'Espace Michelin', and the equivalent for IBM and Ford. This understanding of economics and economic space in terms of power and the explanation of the degree to which the location of distribution of economic activity is not value-free, but reflects choices that do not arise from the benign operation of a non-political economic system, makes for an interesting cartography. A blunter account was offered by the map series *Nuclear Weapons Accidents: The Military-Industrial Atlas of the United States* (Philadelphia, 1978; and 1982) produced by National Action/Research on the Military Industrial Complex. On a different scale, an individual factory or agricultural area involves structures and flows of power that can be presented and, in part, understood in spatial terms, and the social resonance of this approach can be widened if the analysis is extended to consider the accompanying residential patterns. Work and wealth move around the system, creating patterns, and also leaving traces such as the arrangement and layout of buildings, both individual structures and larger groups.

More generally, the information conveyed in a map can be given different spins or resonances by the presentation of the map, particularly the title, caption and symbolization. Thus, for example, a map of modern atomic power plant in a country could be presented in a 'neutral' fashion by simply depicting the location of the plant and associated power grid lines. The symbols could be low key, for example small circles or dots in neutral colours, the title could be 'The Nuclear Industry in . . .', and the caption refer to the quantity of electricity thus produced.

By contrast, the very same information could be presented differently by the use of threatening symbols, for example large pulsing circles in glowing colours, the use of danger colours, especially red; the title could be 'The Threat from Nuclear Power Plants'; and the caption could refer to the danger of contamination and accidents. This last could be taken a stage further if the map also depicted, first, the routes taken by nuclear supplies and products, routes shown as crossing much of the country, secondly, the sites of contamination, leaks and spillages, and, thirdly, the areas of a country within 100 miles of a nuclear plant and prevailing wind directions. Thus the nuclear industry ceases to be specific in its presence and impact. This can also be given an international dimension by showing how routes, spillages and areas of possible contamination cross, or can cross, borders. Thus, the nuclear industry in the USA suddenly becomes a central issue for Canada. The entire effect could be completed by choosing an accompanying picture of a devastated Chernobyl rather than a functioning nuclear plant photographed in sunshine and against a background of green hills.

Companies themselves employ maps in order to make statements about their activities and ideals. Obvious instances are maps that depict the processes by which manufacturing particular products, such as cars, draw on raw materials and sub-assembly work from different countries, thus demonstrating that a company spreads value, and is not simply a multinational 'parasite'. Another example is provided by maps that stress the global appeal of products and thus their naturalness: the world is wholesome.

By the same device, products cease to appear foreign. This was highlighted by the cover of the American news-weekly magazine *Time* on 15 May 1950, which showed a globe with facial features eagerly drinking from a bottle of Coca-Cola. Thus, the global range of the product was proclaimed and it was made to appear natural. Readers who drank Coca-Cola were taking part in a worldwide activity, and this was something to be proud of, because it was American: America and the world were linked and as one, to the enjoyment of the latter and the profit of the former. Maps also serve companies in the analysis of supplies, transport, markets and rivals, offering an important visual component in the analysis of such data and in the creation and implementation of marketing strategies.

The conventional maps of economic activity – here be steel mills – also involve choice. First, they reflect a marked preference for production over consumption. Secondly, the production that attracts attention is manufacturing, not service industries or the financial sector. Thirdly, the concentration is on heavy, not light, industry, steel and shipbuilding, not electronics or cosmetics, let alone industrial activity that requires little investment or for which work is non-skilled or part-time. Fourthly, work, rather than ownership, tends to be mapped, although the work is generally treated as unproblematic and unpolitical. For example, trade-union activity is not usually presented alongside work. There are serious data problems affecting trade-union geography, particularly outside the 'First World', but it can still be mapped.[2]

Work is far easier to assess than ownership, although capital can be conceptualized in spatial terms. Furthermore, money has increasingly become a commodity that is independent, traded separately, in greater volume and with greater volatility than goods, and also one that is part of a globally integrated market.[3] International capital markets and flows play a significant role in the success of and failures of political programmes, both international, for example European monetary convergence, and national. Thus it is increasingly important to map flows of money, and also more necessary to have a global spatial sense in order to understand the flows. The global nature of influential modern media networks is also such that they should be mapped and mapped globally. The international character, range and intentions of networks such as those of Rupert Murdoch or CNN excite interest and controversy, and mapping them indicates an awareness of their political role.

The general maps of economic activity included in atlases reflect a conservative definition of economic activity, one that is very misleading in terms of wealth deployed and created or employment produced. In part, this is due to ease of mapping. It is simpler to map an activity that occurs only on a few sites, such as steelmaking, rather than one that is widespread, for example house painting, unless the latter is an aspect of agriculture that can be used in order to colour in a portion of the map: here be wheat, is easier to show than here be bakers, although, in fact, many other crops will be cultivated in areas depicted as used for wheat; as ever, the use of homogeneous blocs is misleading. An emphasis on 'principal crop' can also pose problems, as, aside from the issue of rotation, there is the question of 'principal' in what sense: acreage, financial value, or employment?

It could be argued that there is no point depicting widespread activities, for example banking or do-it-yourself work, as they are so widely distributed that the maps would only correspond to those of towns and people respectively. Ease of mapping is one thing; usefulness of mapping is another. Yet, in terms of wealth deployed and created, it is strange to produce a map of the American economy that includes the Texan petrochemical industry but not Wall Street. Hollywood is also generally excluded.

More generally, the reality of the wealth and consumption dynamics of socio-economic systems have to be captured by mapping that reflects shifts in activity. Whereas it is and was useful to map landholding patterns for, say, interwar Poland, land is now far less of a measure of wealth in most societies, especially in the First World. Yet, by their very nature, it may be difficult to spot emerging sectors or to map them adequately.

The data available for economic mapping are skewed. As ever, there is much more for the 'First' than the 'Third' World, and that for the latter concentrates on economic sectors defined by Western activity, for example large-scale mining, forestry, cash crops and the production of goods for Western markets, and the data are commonly presented with reference to Western categorizations. This situation has altered relatively little since decolonization, especially in the poorer parts of the world, where resources do not exist for extensive new mapping projects. Instead, cartographic agendas and activity remain greatly affected by Western patterns, not least Western economic interests.

Despite limitations with the data, some atlases, nevertheless, map a wide and interesting range of economic indicators. Michael Kidron and Ronald Segal's *State of the World Atlas* (5th edn, London, 1995) includes a spread on 'Value Subtracted'. This maps 'Relative Exploitation. Payment received by workers in proportion to wealth they produce.' Somewhat surprisingly, Niger and Rwanda top the percentages, and the peak group includes such workers' paradises as China, Brazil, El Salvador and Guatemala, all apparently less exploitative of workers than states such as Bolivia, Japan, Nigeria, Somalia and the USA, which, in turn, are less exploitative than the Netherlands, Norway, Canada and Iran, which are all less exploitative than France. As the supporting text notes, exploitation exists, but the map offers a crude and, arguably, misleading account. This spread can be seen as indicative of the general tone of an atlas that in its Introduction attacks business culture and compares it to organized crime. There is also a very critical spread on privatization.

Where economic data does exist, the growing number of mathematical techniques that can be applied permit the analysis and depiction of more power relationships than used to be the case. For example, the degree to which a region is integrated into more widespread economic links reflects its vulnerability to such relationships. Mathematical indices can be mapped, so that, to take the case of a map of the Scottish grain market,

> the basis of this map is the idea that if a matrix of correlation coefficients is built up comparing each county's short-term price movements with those of every other county, and if such coefficients can be used as an approximate measure of market association, then by aggregating each county's correlation coefficients one has, in effect, a relative measure of the degree to which short-term price movements in that county relate to those taking place throughout the rest of the country.

The nature of economic space is itself subject to debate.[4] Classically, economic theory was interpreted spatially in terms of isotropic (unvarying, uniform) surfaces, which offered models that geographers could consider in theories of locational analysis. Such theories and models now, however, look somewhat mechanistic. It is also clear that *cultures* of economic activity do not readily lend themselves to mapping, as the location of manufacturing plant does, although patterns of investment and consumption can indeed be mapped. Arguments that cultural geography played and play a major role in economic life, for example Terry Jordan's account of different ranching traditions, their pedigrees and impact in his *North American Cattle-Ranching Frontiers: Origins, Diffusion and Differentiation* (Albuquerque, 1993), pose problems for mapping, as, more generally, do questions of, for example, the relationship between cultural indicators, such as Protestantism, and economic life.

Social issues

Insofar as they can be separated, social issues are harder and more contentious to map than their economic counterparts, and the choice of topics and nature of the mapping can be seen as political. The notion of social space is a politicizing concept, for it presents the construction of space as an aspect of social policy that is not value-free and that is likely to be controversial. The French sociologist Henri Lefebvre has argued that every society has developed a particular social space that matches its economic and social needs.[5]

However, there are major problems in mapping social patterns. These have increased as a result of a widespread reaction against the positivism of much spatial science. A positivist viewpoint aids mapping, as it minimizes the role of human agency in favour of a structuralist interpretation in which human choices are determined, or at least heavily influenced, by parameters, some of which can be mapped. Thus, for example, settlement patterns can be seen as a consequence of environmental factors or of the nature of market systems. By contrast, an emphasis on humans as the crucial element in geographical change and patterns, on concepts and ideologies, on the role of human decisions, leads to a less clear-cut and more 'messy' situation, especially if the individual rather than the collective is seen as the basic unit of decision-making. Patterns become less apparent and processes of causation less easy to define or to agree upon, and, as a consequence, the relationships that should be displayed in maps and models are less obvious.[6]

Other problems arose from a different understanding of the nature of social organization. A class-based analysis of society and politics lent itself to mapping, provided that class was seen in relatively simplistic terms and with reference to factors of production. As analyses of class have become more complex, so the possibilities for mapping have become more problematic.

Although there is a parallelism between geographical and social difference,[7] only the most coarse-grained social divisions can be mapped in any readily interpretable way. When social divisions were *fairly* basic and clear, it was possible to map them. This was true, for example, of many cities in the early twentieth century; they were societies and polities divided by immigrant/ethnic status and by class.

These divisions can be shown, although such a process involved and involves problems, not least the relationship between class and ethnicity. Under close scrutiny, complex residential patterns emerge, and it is apparent that most of the urban landscape was and is not composed of neighbourhoods made homogeneous by ethnicity, class or a combination of the two. Nevertheless, valuable differences can be depicted. In their *Atlas of Housing Conditions in Welsh Districts* (Swansea, 1988), R.C. Prentice and G.B. Lewis sought 'to provide in a readily accessible form information, firstly, on housing quality and tenure in Wales, and, secondly, on relevant demographics and deprivation as a background to Welsh housing needs' (p. 2). Derived from information in the 1986 Welsh Inter Censa Survey, such as percentage of renters in receipt of rent rebate, the maps revealed a clear diversity of conditions. It is easy to criticize the problems of using the data, not least the extent to which the units employed sometimes served to encompass very different circumstances. However, the atlas served to demonstrate a complexity that had consequences for planning,

Because of this diversity it is important to recognise that both explicit and implicit housing policies will likely have different impacts across Wales, frequently without spatial consistency. In forming and appraising housing policies in Wales it is important to anticipate this diversity and not implicitly to assume either a uniform impact or a spatially cohesive impact.

More generally, even if scholars can integrate both housing and labour markets, each of which can be seen as political in cause and consequence, in order to understand and map residential patterns,[8] there are other 'big' divisions and issues in society that have to be considered: the obvious case is gender. Certain issues related to gender can be very difficult to map. Male/female proportions can be mapped, but they offer little guidance to sexual politics. A more promising approach, which emphasizes spatial considerations and the notion of control, has been that of approaching the body, the home, the street and the community as political territories, structured by practices and discourses of power.[9]

However, by assuming that conflict and control are the norm in gender (and more generally social, economic and political) relationships, such an approach risks using cartographical methods, forms and languages to give credence to and amplify a set of assumptions that are questionable, or, at least, require demonstration. This, indeed, is a common problem with the use of maps. Cartographical depiction rests on assumptions about relationships, both specifically, in spatial terms, and more generally, and yet many of these assumptions proceed from a priori views. This is, of course, true of maps from the technical standpoint. Assumptions about scale, equivalence, projection and meaning underlie the production and reading of maps. To suggest, therefore, that the spatial encoding of social ideologies, such as environmental determinism or the notion that society is a realm of conflict, poses a problem because it simply demonstrates assumptions, thus inviting the rejoinder that that is the nature of mapping. Yet the scale of a map or its graticules are devices, their precision necessary for the document's use. By contrast, a map of say Britain in terms of spouse-abuse is a statement, especially if no other aspects of gender is mapped, that such abuse is crucial to gender relations and that the very visual statement is necessary even if spatial variations in incidence are difficult to assess. Such an approach may be queried. All maps are models informed by theory and ideology, and any assumption of scientific objectivity is dangerous, but such a subjectivity does not imply that anything goes.

One problem in mapping gender is that of scale. Study of gender dynamics and tensions in a spatial context tend to focus on the organization of household and communal space,[10] rather than on the national or international scale. Nevertheless, *The Women's Atlas of the United States* (2nd edn, New York, 1995) by Timothy and Cathy Fast, sought to map the

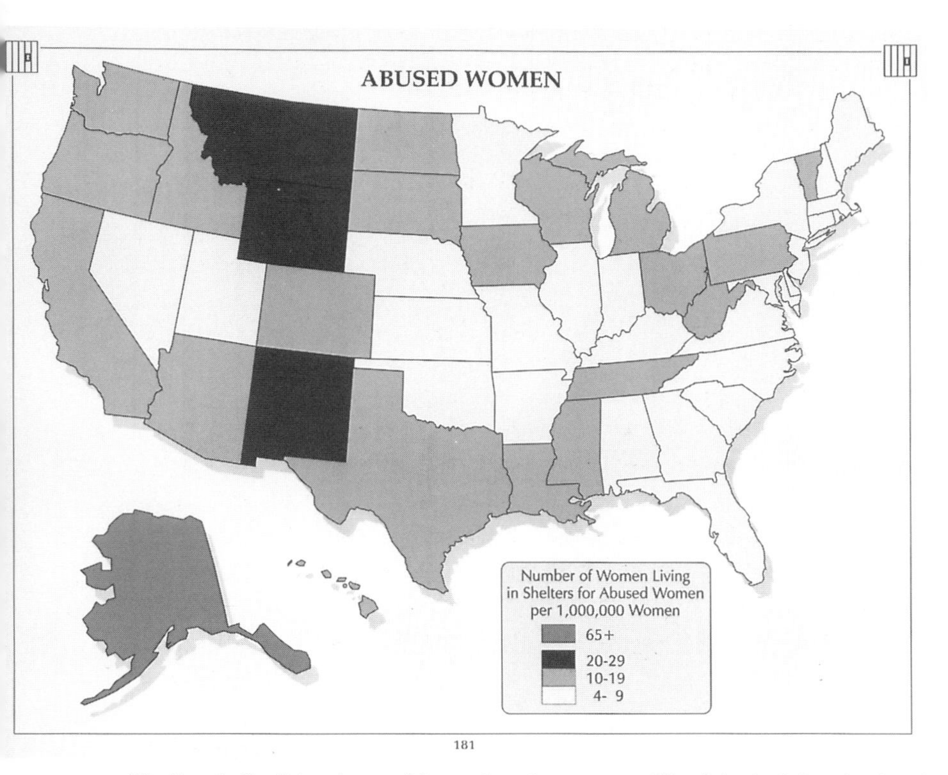

The Female Condition. A map of the number of women per million living in shelters for abused women, from Timothy Fast and Cathy Carroll Fast, *Women's Atlas of the United States* (1995). The use of States of the Union as data units creates serious weaknesses for they amalgamate very different areas.

female experience at the national level. It did so in a campaigning and interactive, almost self-help, fashion:

> Although the atlas will supply specific information on the geography of women, our hope is that it will also open the door to further study of the issues examined here. Why do women in Alaska have such a low incidence of heart disease ... You, of course, should formulate your own questions about the geography of women ... These maps show the differing levels of participation by women from state to state and the ground yet to be made up before true equality is achieved.

The tone of the atlas was clear-cut. Aside from the expression of preference for Clinton over Reagan, there was a clear-cut sense that problems existed and should be confronted. Use of states of the Union as data units created serious weaknesses, for they amalgamate very different areas, such as New York City and the rest of the state, but the authors properly drew attention to some of the deficiencies in the maps. For that of the number of women living in shelters for abused women per million women they noted,

A community's concern for abused women and its willingness to make a financial commitment to the operation of shelters for victims of abuse obviously greatly affects the number of women who are sheltered . . . Religious and cultural factors also influence the number of women residing in shelters. Areas with traditionally high religious affiliations, for instance the Bible Belt, might have a lower incidence of spousal abuse, or more likely a lower incidence of reported spousal abuse. Women in these regions may be less likely to reveal that they are in an abusive relationship because religious teachings place men at the physical and spiritual head of the family.[11]

At the international level, Joni Seager and Ann Olson produced the *Women in the World International Atlas* (London, 1986). Gender issues have also come to play a role in some non-specialist global atlases. Michael Kidron and Ronald Segal's *The State of the World Atlas* (5th edn, London, 1995), a radical work given wide readership by its publication in paperback and by Penguin, includes three relevant spreads. The first, headed 'Missing Women', maps the difference between the number of females in a population and the number who would be alive if they retained their natural female-to-male ratio at birth. The map carries the statement that this registers the difference 'compared with the number that might have lived had they enjoyed the same treatment they get in states of high human development', while the supporting text explains that in China many female babies are killed at birth, while in India many women are overworked, undernourished and neglected. The map also carries a white wraith-like symbol of a woman for countries where there is a 'deficit of more than 500,000 women'. There is a failure to explain the role of differential longevity in accounting for female 'surplus' in states such as Britain; possibly because this does not accord with the message of male oppression provided in the text. There is also a spread on women in work. An inset map, based on UNESCO data, maps the ratio of women-to-men students in higher (including vocational) education.

The following spread, 'Body Politics', carries a main map on the availability of legally available abortion in 1991. The supporting text correctly warns that law does not everywhere reflect clinical practice, but also includes an unsupported and tendentious assertion of the relationship between labour supply and abortion, a relationship that the map does not support:

sometimes when labour is scarce, as in the rich states in the 1960s, women find it possible, although never easy, to extend the frontiers of control over their bodies. In times of recession and idleness, women find such frontiers contracting.

Overleaf A different agenda for global atlases: 'Missing Women' from Kidron and Segal's *State of the World Atlas*. It maps the difference between the number of females in a population and the number who would be alive if they retained their natural female-to-male ratio at birth.

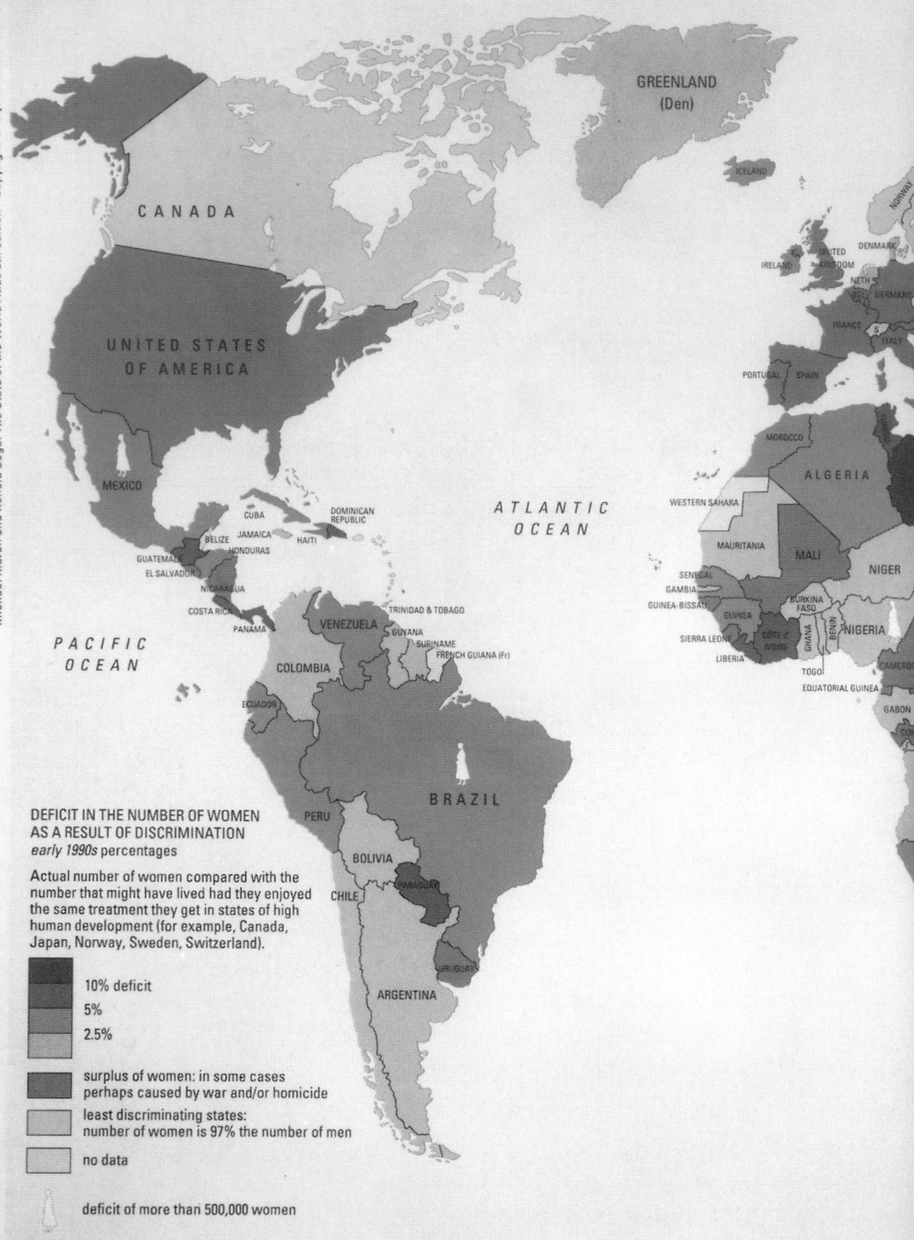

DEFICIT IN THE NUMBER OF WOMEN AS A RESULT OF DISCRIMINATION
early 1990s percentages

Actual number of women compared with the number that might have lived had they enjoyed the same treatment they get in states of high human development (for example, Canada, Japan, Norway, Sweden, Switzerland).

- 10% deficit
- 5%
- 2.5%
- surplus of women: in some cases perhaps caused by war and/or homicide
- least discriminating states: number of women is 97% the number of men
- no data
- deficit of more than 500,000 women

Sources: UN; U.S. State Department.

MISSING WOMEN 16

If women worldwide enjoyed the same treatment as they do in Canada, Japan, Norway, Sweden and Switzerland, there would be 120 million more women alive. If they enjoyed fully-equal treatment with men, there would be many tens of millions more.

In 1992, there were 20 million fewer women than men in China.

RATIO OF WOMEN TO MEN STUDENTS IN HIGHER (INCLUDING VOCATIONAL) EDUCATION *early 1990s or latest available data*

- 25 women to every 100 men
- 50
- more women than men
- no data

Source: UNESCO.

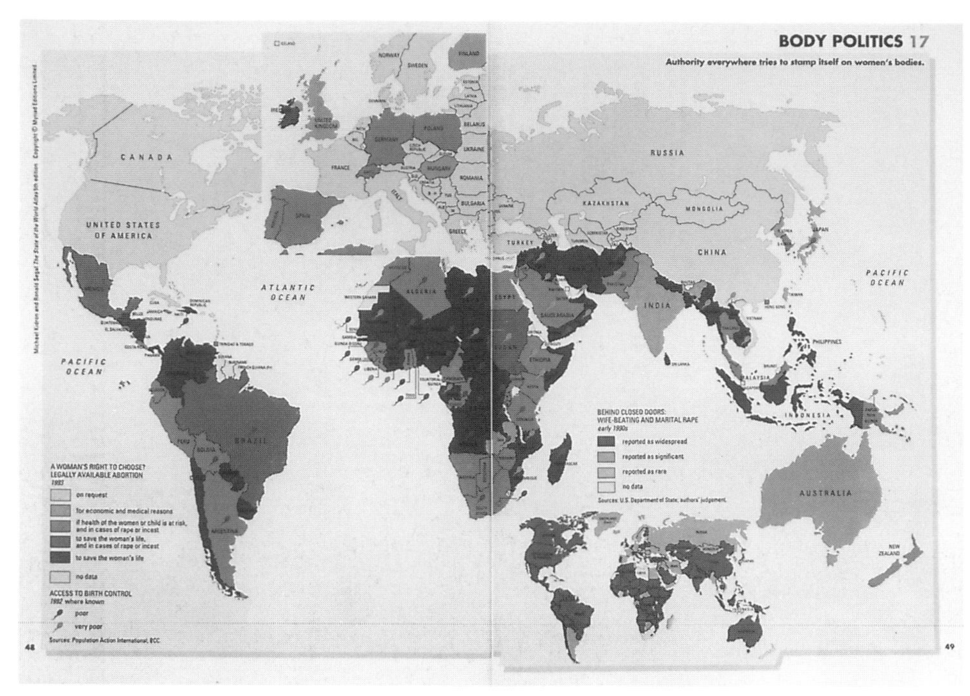

'Body Politics' from Kidron and Segal's *State of the World Atlas*. A map on the availability of legally available abortion is accompanied by an unsupported assertion of its relationship to labour supply.

> Nowhere is this more apparent than in Eastern Europe where the collapse of Communism, profligate with labour as with so many other things, has meant a sharp turnabout in what were once among the most liberal abortion regimes in the world. The personal has paid heavily for the political in those countries.

This assertion is in no way borne out by the map and is anyway a very partial account of what is a far more complex situation. The role of religion, for example, is ignored.

The inset map on the spread also contains some surprises. Entitled 'Behind Closed Doors: Wife-Beating and Marital Rape Early 1990s', it gives as sources 'US Department of State; authors' judgement', and includes Russia, Ukraine and Iran as countries where these are 'reported as significant', while the more serious category, reported as 'widespread', includes Britain, Canada, Finland and the USA. 'Reported as rare' is a small group including Norway, Ivory Coast, Laos and Madagascar. The supporting text notes that 'records are haphazard and scarcely scientific', which makes the decision to produce the map surprising, not least as widespread, significant and rare are particularly subjective criteria and are not explained.

As a separate issue, there appear to be important differences between men and women in their reading of maps. For example, recent research has indicated that men are much more likely to orient their maps with north at the top, while women are more likely to use a variety of line symbols.[12] Given the emphasis on the role of the map-reader in interpreting maps, indeed in appropriating them to pre-existing views, it is clearly important that such research be pursued further.

Mapping social divisions is especially difficult in the modern world, because the multiple nature of identities has become more of an issue. Aside from the multiple identities we all feel (class, gender, sexual identity, age, family status, occupation, consumption patterns, location, country, school and so on), these identities exert and express themselves differently in each person. Further, the various identities individuals hold overlap, and their influence varies by issue, circumstances and stage of life. For example, one person may identify with others of their ethnic background when that seems relevant and forget about it when not. It is not therefore clear what a map of Turks in Germany or Jews in the USA is supposed to communicate. In addition, if it true that 'in our mobile society, people are continually forced to occupy, and to adjust to, brand-new environments',[13] these shifts themselves refract and are refracted by identities.

Identity is clearly not 'equally fluid for all'. To map the environmental perceptions of, for example, wheelchair users in Coventry is to offer a valuable reminder of yet another range of spatial response and accessibility, but of course such users vary in their concerns and resources. To treat West Hollywood as the setting for a homosexual identity, as indeed a gay townscape, is to present an understanding of the place that may not only be controversial, but that may underrate different attitudes among homosexuals. Sexual preferences are frequently treated like ethnicities: reified, homogenized and made a causal variable. This may be very misleading. More generally, the 'geography of sex' poses problems of cartographic analysis and depiction that involve politics because of attempts to politicize issues of sexual identity, identification and activity.[14]

It is far from clear how to map a maelstrom of overlapping and potentially conflicting divisions, but they are of the essence of the complex nature of modern society and politics. The obvious strategy is to focus on what seems most meaningful and mappable, and to recognize the resulting bias. The complexity of the relationship between space and society is such that the limits of what maps can convey as analytical texts are reached quite quickly.

In practice, cartographers cannot personally survey large countries. If they want to map a new subject, or an old subject by a new method, they have two sources of information, both outside their own control. The first is large-scale topographical maps. These generally contain very little economic, social and cultural information, although both the maps and their contents reflect economic, social and cultural values. The second is lists and tables that include a locational component. Most of these are compiled either by governments or by well-funded commercial or professional associations. Cartographers are seldom asked to suggest what topics should be covered by these sources; in short, they have little influence over the nature of their data set.

Furthermore, the large proportion of modern statistical information that is based on sampling is an important practical and theoretical issue. A sample may be just large enough for national use. But a part of that sample will – almost by definition – be too small for use in a part of the national territory. In other words, many modern national statistics are unmappable or, if mapped, untrustworthy. In the field of information-gathering, it is the censuses that lead and the maps that follow. If a topic has never been surveyed in a statistical sense this may only be an accident, but it may reflect the nature of the subject. Some subjects cannot be mapped in terms of modern objective mapping because they cannot be tabulated.

The real division would then be between what can be mapped in such terms or listed (and written about), and what can only be written about. It is unhelpful for writers of the ideological school, such as Harley, to blame cartographers for not mapping the unmappable. It is clear that maps and verbal texts may be in a sense 'continuous', and that expressions in one medium cannot be considered in isolation from those in the other. But this does not imply that maps can do everything that words can do, or that the map and its cartographer should be found wanting if they fail to do so. The limitations of cartography, whether in conventional modern Western terms or more generally, mean that it is necessary to know whether a commentator is referring to maps or texts or a mixture of them both.

At both the macro and the micro level, there are also 'small' divisions that are, nevertheless, very meaningful for individuals, for example the difference between a first-, second- and third-generation migrant from the countryside in, say, Nairobi, or between an affluent and a poor inhabitant of a British village, the car-less latter confined spatially to a degree that would not make sense to the former. It is unclear how these finely grained aspects of society can ever be mapped, except, perhaps, on a block-by-block basis. Furthermore, even if that scale of mapping is available, there is a problem with obtaining or collecting sufficiently detailed information.

There are readily apparent problems with different systems of mappable classification of social structure, and thus with mapping social politics.

Wealth or occupational status might appear clear-cut, but each poses problems. The measurement and thus mapping of wealth is affected by issues such as the convertibility of assets, the value of assets that are yet to mature, tax regulations, accounting practices, financial commitments, and the accuracy of returns. There are different types of wealth, and these can affect identities and political responses; different types, however, also tend to have locational characteristics that are readily mapped. The benefit of refining the classification and mapping of wealth is indicated in John Mollenkopf's *New York City in the 1980s: A Social, Economic and Political Atlas* (New York, 1993). Mollenkopf used the 1980 city's 2,200 census tracts, each of which contained about 3,300 people. His map of 'Change in Median Household Income, 1980–1990' demonstrated changing patterns of income inequality, but that was shown even more clearly in the map of changes in 'unearned' income: income from dividends, interest and rent in the same period. Much of the 1980s boom took the form of higher returns on such assets, rather than wages, and Mollenkopf's map indicated that, measured in financial terms, the benefits of the decade were concentrated in White upper-middle-class areas; indeed, these areas could, in part, be defined by reference to such criteria. Similarly, in Nantes, publicly available records of income-tax return have made it possible to draw the fiscal contours of an urban environment with great precision. They also permit the display of social dynamics by adding the perspective of time.[15]

Occupational distribution also poses problems for the cartographer, not least of classification and categorization. There are major differences within whichever strata are employed, and occupational groupings do not always reflect those of status or wealth.

There are other major difficulties in the mapping of social indicators. First, much of the work on social and political anthropology has been on the local scale, and it is difficult to see how an atlas treatment on a national or international scale can be made when so much of the available material is only local studies. Secondly, there is the problem of the socio-political response to objective data. Nutrition or housing or educational and health provision can be mapped, but what the figures and the spatial distribution represent to individuals, communities and commentators, past, present and future, varies greatly. The contingent nature and interactions of relative and 'absolute' values are difficult to map. This is even more the case if the temporal dimension is added. A map of household overcrowding in Newcastle in 1971 carried with it the warning that

> apart from the problems of precise measurement, standards and expectations are constantly rising, and this makes overcrowding a rather nebulous and diffi-

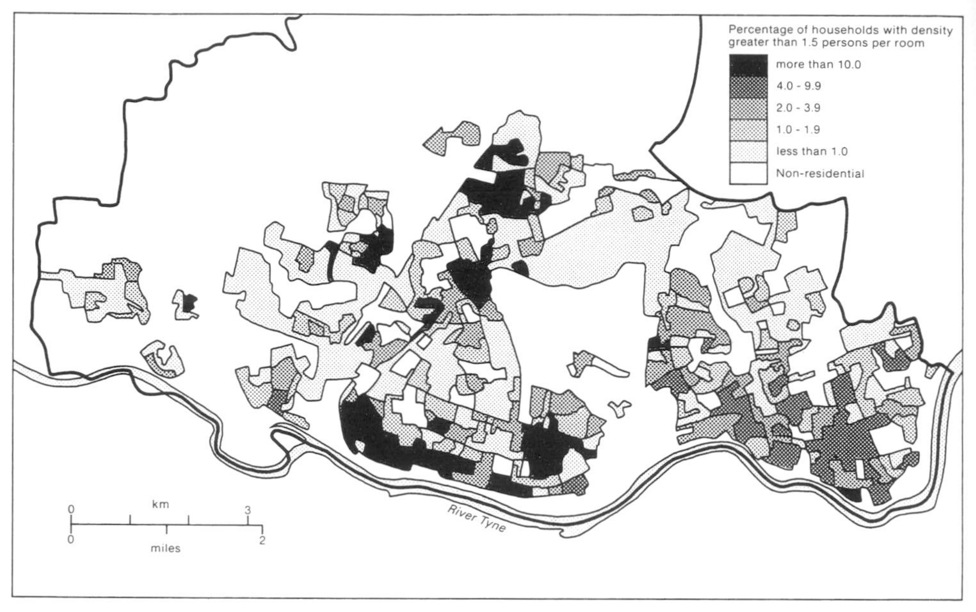

Overcrowded households in Newcastle, 1971, from M. Barke and R. J. Buswell's *Newcastle's Changing Map* (1992). As the text notes, aside from the problems of precise measurement, rising standards and expectations make overcrowding difficult to define. Nevertheless, the contrast between the West End and Gosforth, north of the Town Moor, is readily apparent.

cult concept to define. For example, not too long ago a density of 2.0 persons per room was considered to be the semi-official definition of overcrowding but in the post-war years this has been reduced to a density of over 1.5 persons per room.[16]

Thirdly, there is the question of how best to portray the landscape impact of social and ethnic groups, and how to present some sense of their activity and visibility. Parcelling residential areas into discrete 'communities', in the style adapted from the traditional urban sociology of the Chicago School, helps the cartographer, but it has serious problems, not least because the patchwork-quilt effect it creates suggests to the user clashing regions of homogeneity analogous to the areas of uniform sovereignty separated by sharply demarcated boundaries seen in political cartography.

Instead, it is possible to focus on public space, although that also creates problems, and can, again, lead to an emphasis on clashes. It is feasible to observe who claims rights to the use of public space, the response of the population in general, and the exercise of control by municipalities and their agents. The strategy captures 'visible' social groups, that is, those which risk a public presence – and those which know they could present themselves with approval. In short, it is evidence of politicization, of political pretensions and political hierarchies. Clearly, some categories of people may remain invisible if cartographers adopt this approach; indeed, it is part of the politics of space that some groups are marginalized.

However, in most societies there is a considerable variety of groups willing to act or demonstrate in public, sometimes as invited participants in governmental or civic celebrations, but more often on their own initiative. The action or demonstration may create the space. Social and ethnic distinctions were and are richly mirrored in the use of public space for collective action and demonstration. Orange Order marches in Northern Ireland or processions in Israel are an obvious example: space is being defined and used in a political fashion, and disputes over where marches can proceed reflect highly charged senses of control over space and of the role of spatial power in ethnic identity. Cartographically, however, it is difficult to capture a moving spectacle. The dynamism of a march or riot, the enthusiasm or respect shown by participants (marchers and onlookers), are qualities that are integral to the occasion, but that are difficult to convey.

To put to one side the issue of movement, it is, nevertheless, the case that contention over the public character of space can capture an important aspect of the dynamics of society. Thus a map of roads or road schemes can be complemented by another concentrating on attempts to resist such schemes. This can be presented in terms of an 'objective' onlooker or, alternatively, the spatial perceptions of protesters can be mapped, both in terms of the particular site where they seek to resist new or wider roads, for example in Britain in the 1990s near Bath, Newbury, Winchester or Salisbury, and more generally. In the former case, it is necessary for the map to capture vistas, to reveal the impact of sound, and to hint at the aesthetic and spiritual power and visual range of trees. More widely, in *The Power of Place. Urban Landscapes as Public History* (Cambridge, MA, 1995), Dolores Hayden has emphasized the power of ordinary urban landscapes to develop and sustain citizens' public memory, thus ensuring that place is linked to identity and therefore political power and purpose.

The mapping of social issues is, therefore, potentially very political. It subverts any emphasis on homogeneity produced by the mapping of sovereign authority and presents societies as complex and divided. Shifts within medical mapping, for example, reflect the politics of mapping social themes, as epidemiological mappers have scrutinized differently the socio-political contexts and configurations of epidemics. The mapping of the social dimension also opens up the possibility of a radical cartography. At the level of open defiance, the mapping of revolution and riot is a reminder of the contingent nature of sovereign authority and the controverted character of sovereign power. By directing attention to issues of social distribution, mapping can also open the politically charged question of social justice.

Radical cartography can be addressed both in global terms and with reference to individual countries. In both, the distribution of resources and problems can be mapped, inviting questions about inegalitarianism. Such questions are made more pointed by the role of democratization in the twentieth century and the nature of democratic ideologies. Any cultural stress on equal rights and opportunities invites a cartographic questioning, and makes the spatial distribution of such rights and opportunities a ready issue. Maps are the best way to highlight spatial differences visually, and thus the culture of democracy makes mapping by its nature potentially subversive.

Radical cartography also offers the possibility of problematizing generally accepted notions of progress. Thus, the *Third World Atlas* (Milton Keynes, 1983), prepared by Ben Crow and Alan Thomas, presents its spread on industrialization as 'The Destruction of Handicrafts and the Rise of Machinofacture', while the map on railways carries with it a note explaining communications in terms of power: 'Note that the pattern of railways generally serves seaborne, international trade rather than intra-regional trade'.

Atlases for the environment

The environment is another sphere for radical cartography, an interesting reversal of a traditional cartographic theme, because environmentalism – a somewhat deterministic view of the impact of the environment on human development and activity – was influential in cartography, especially in the first half of the twentieth century, and was commonly presented as unproblematic. It explained locations and boundaries and thus made them appear natural.

By the end of the century, however, the human impact on the environment was a major item of concern for atlases. Pollution and environmental degradation have not only been mapped in specialist works; they have also become an issue in mainstream cartography. Discussion of pollution has led to a tempering of the teleological progressivism that had long affected the text in atlases and, indeed, been a theme in mapping for which more equals best: for long, maps had concentrated on human activity within an intellectual context that saw such activity as beneficial, an improvement of the world. Thus, to map more towns and roads, or the expansion of the cultivated area, was to reveal the march of progress, a march that also brought more data. The widely circulated American *National Geographic* magazine for long praised economic development and equated the West with progress and civilization. *The March of Civilization in Maps and Pictures* (New York, 1950) declared of the USA 'its freedom-loving people have devoted their energies to developing the riches that Nature has so lavishly supplied'.

By contrast, the *American Heritage Pictorial Atlas of United States History* (New York, 1966) stated,

> technological change, the population explosion, and the unchecked march of urban civilisation have upset the balance of nature and created new environmental problems, such as those posed by impure air and polluted streams, problems unimagined by the pioneers who settled the nation.[17]

The London-based *Geographical Magazine* of January 1973 printed two articles on Amazonia, each supported by a map and written by a geographer. One, by Edward Leahy, on the Trans-Amazonica Highway, mapped roads, existing, under construction and projected, rivers and state boundaries. The text included the following:

> One is moved with admiration at the manner and spirit of the Trans- Amazonica undertaking. Young Brazilians, bursting with optimism and trained and equipped with the most sophisticated paraphernalia of modern civilization, are in direct confrontation with the jungle. The road under construction in the Amazon basin will do much to bring about the taming of this great wilderness bastion.

The subsequent piece, by Edwin Brooks, concentrated on the problems facing the Brazilian Indians. The article was more subtle and searching, and the map – of the author's travels – recorded major Indian parks and reservations. It was amplified by two inset maps, one of major new roads and the other of Indian cultural areas, although the latter were presented within a non-Indian frame, not least because the map was shaped by alien entities and interpretative schemes: the areas were organized with reference to the Brazilian frontier, the Equator and the Tropic of Capricorn.

The *Atlas of Florida* (Gainesville, 1992) referred to the Everglades:

> Engineering works were undertaken to deal with the perceived excess of water in the region, but more recently potential long-run deficiencies in South Florida's water supply have begun to be recognized. Problems of sound land and water management in South Florida remain unsolved ... By 1990 thirteen million Floridans had a far less stable relationship with the environment than had the Indians of Ponce de Léon's time ... Excessive drainage of wetlands, construction of miles of artificial waterfront, hazardous waste discharges, and unplanned urban sprawl are manifestations of population growth outrunning orderly, careful accommodation to the special qualities of the Florida environment.[18]

More explicit, environmental mapping tends to be more aggressive politically. Thus the commitment, energy, imagination and arresting graphics of *The Gaia Atlas of Planet Management* (London, 1985) was somewhat dated by simplistic flashes, as when David Bellamy, in the Fore-

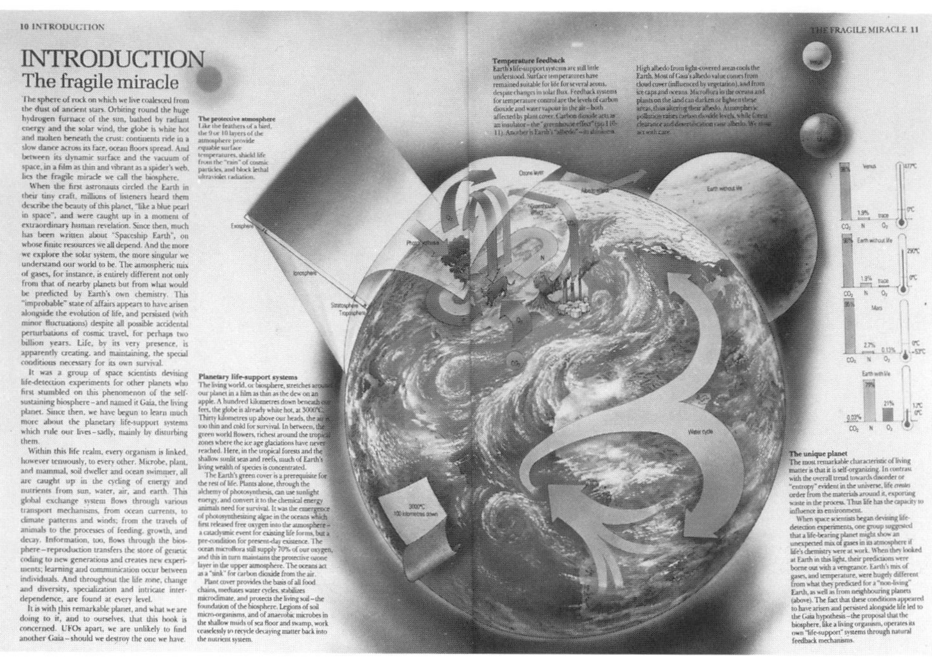

10 INTRODUCTION

INTRODUCTION
The fragile miracle

The sphere of rock on which we live coalesced from the dust of ancient stars. Orbiting round the huge hydrogen furnace of the sun, bathed by radiant energy and the solar wind, the globe is white hot and molten beneath the crust: continents ride in a slow dance across its face, ocean floors spread. And between its dynamic surface and the vacuum of space, in a film as thin and vibrant as a spider's web, lies the fragile miracle we call the biosphere.

When the first astronauts circled the Earth in their tiny craft, millions of listeners heard them describe the beauty of this planet, "like a blue pearl in space", and were caught up in a moment of extraordinary human revelation. Since then, much has been written about "Spaceship Earth", on whose finite resources we all depend. And the more we explore the solar system, the more singular we understand our world to be. The atmospheric mix of gases, for instance, is entirely different not only from that of nearby planets but from what would be predicted by Earth's own chemistry. This "improbable" state of affairs appears to have arisen alongside the evolution of life, and persisted (with minor fluctuations) despite all possible accidental perturbations of cosmic travel, for perhaps two billion years. Life, by its very presence, is apparently creating, and maintaining, the special conditions necessary for its own survival.

It was a group of space scientists devising life-detection experiments for other planets who first stumbled on this phenomenon of the self-sustaining biosphere – and named it Gaia, the living planet. Since then, we have begun to learn much more about the planetary life-support systems which rule our lives – sadly, mainly by disturbing them.

Within this life realm, every organism is linked, however tenuously, to every other. Microbe, plant, and mammal, soil dweller and ocean swimmer, all are caught up in the cycling of energy and nutrients from sun, water, air, and earth. This global exchange system flows through various transport mechanisms, from ocean currents, to climate patterns and winds; from the travels of animals to the processes of feeding, growth, and decay. Information, too, flows through the biosphere – reproduction transfers the store of genetic coding to new generations and creates new experiments; learning and communication occur between individuals. And throughout the life zone, change and diversity, specialization and intricate interdependence, are found at every level.

It is with this remarkable planet, and what we are doing to it, and to ourselves, that this book is concerned. UFOs apart, we are unlikely to find another Gaia – should we destroy the one we have.

THE FRAGILE MIRACLE 11

'The Fragile Miracle' from *The Gaia Atlas of Planet Management* (1994). A presentation of the globe as an environmental system affected by human activities such as atmospheric pollution.

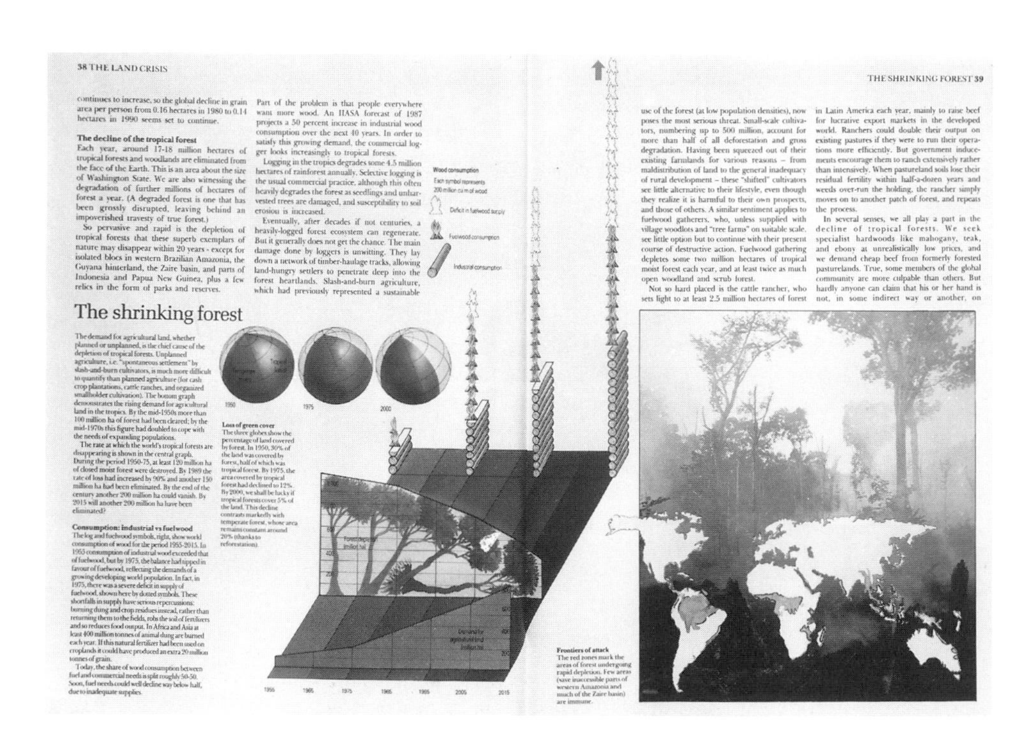

38 THE LAND CRISIS

continues to increase, so the global decline in grain area per person from 0.16 hectares in 1980 to 0.14 hectares in 1990 seems set to continue.

The decline of the tropical forest

Each year, around 17-18 million hectares of tropical forests and woodlands are eliminated from the face of the Earth. This is an area about the size of Washington State. We are also witnessing the degradation of further millions of hectares of forest a year. (A degraded forest is one that has been grossly disrupted, leaving behind an impoverished travesty of true forest.)

So pervasive and rapid is the depletion of tropical forests that these superb exemplars of nature may disappear within 20 years - except for isolated blocs in western Brazilian Amazonia, the Guyana hinterland, the Zaire basin, and parts of Indonesia and Papua New Guinea, plus a few relics in the form of parks and reserves.

Part of the problem is that people everywhere want more wood. An IIASA forecast of 1987 projects a 50 percent increase in industrial wood consumption over the next 40 years. In order to satisfy this growing demand, the commercial logger looks increasingly to tropical forests.

Logging in the tropics degrades some 4.5 million hectares of rainforest annually. Selective logging is the usual commercial practice, although this often heavily degrades the forest as seedlings and unharvested trees are damaged, and susceptibility to soil erosion is increased.

Eventually, after decades if not centuries, a heavily-logged forest ecosystem can regenerate. But it generally does not get the chance. The main damage done by loggers is unwitting. They lay down a network of timber-haulage tracks, allowing land-hungry settlers to penetrate deep into the forest heartlands. Slash-and-burn agriculture, which had previously represented a sustainable use of the forest (at low population densities), now poses the most serious threat. Small-scale cultivators, numbering up to 500 million, account for more than half of all deforestation and gross degradation. Having been squeezed out of their existing farmlands for various reasons – from maldistribution of land to the general inadequacy of rural development – these "shifted" cultivators see little alternative to their lifestyle, even though they realize it is harmful to their own prospects, and those of others. A similar sentiment applies to fuelwood gatherers, who, unless supplied with village woodlots and "tree farms" on suitable scale, see little option but to continue with their present course of destructive action. Fuelwood gathering depletes some two million hectares of tropical moist forest each year, and at least twice as much open woodland and scrub forest.

Not so hard placed is the cattle rancher, who sets light to at least 2.5 million hectares of forest in Latin America each year, mainly to raise beef for lucrative export markets in the developed world. Ranchers could double their output on existing pastures if they were to run their operations more efficiently. But government inducements encourage them to ranch extensively rather than intensively. When pastureland soils lose their residual fertility within half-a-dozen years and weeds over-run the holding, the rancher simply moves on to another patch of forest, and repeats the process.

In several senses, we all play a part in the decline of tropical forests. We seek specialist hardwoods like mahogany, teak, and ebony at unrealistically low prices, and we demand cheap beef from formerly forested pasturelands. True, some members of the global community are more culpable than others. But hardly anyone can claim that his or her hand is not, in some indirect way or another, on

The shrinking forest

THE SHRINKING FOREST 39

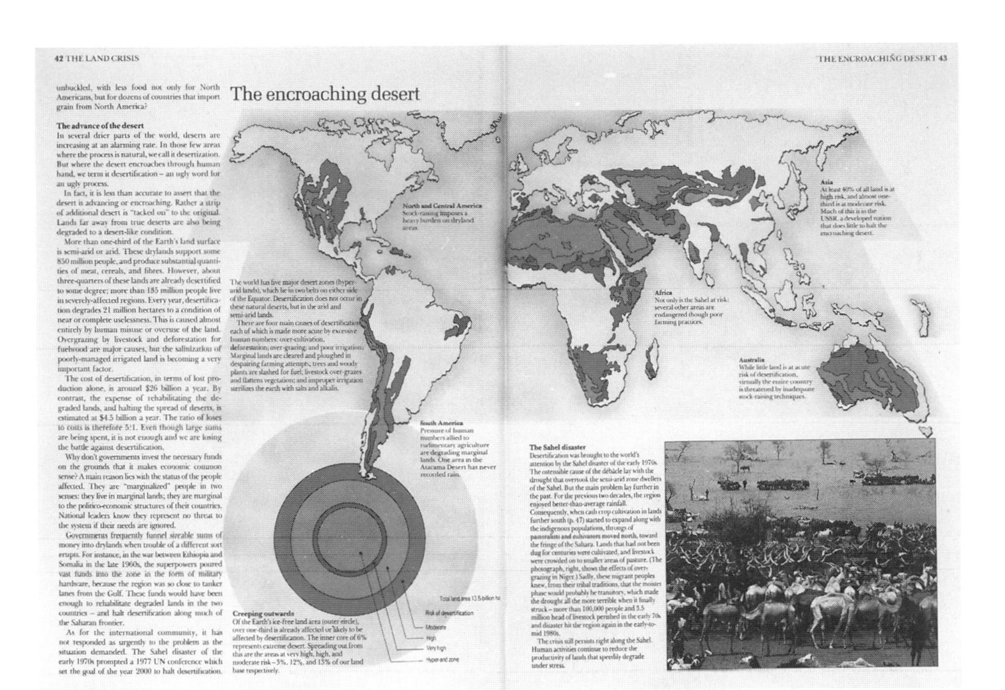

42 THE LAND CRISIS

unbuckled, with less food not only for North Americans, but for dozens of countries that import grain from North America?

The advance of the desert

In several drier parts of the world, deserts are increasing at an alarming rate. In those few areas where the process is natural, we call it desertization. But where the desert encroaches through human hand, we term it desertification – an ugly word for an ugly process.

In fact, it is less than accurate to assert that the desert is advancing or encroaching. Rather a strip of additional desert is "tacked on" to the original. Lands far away from true deserts are also being degraded to a desert-like condition.

More than one-third of the Earth's land surface is semi-arid or arid. These drylands support some 850 million people, and produce substantial quantities of meat, cereals, and fibres. However, about three-quarters of these lands are already desertified to some degree; more than 135 million people live in severely-affected regions. Every year, desertification degrades 21 million hectares to a condition of near or complete uselessness. This is caused almost entirely by human misuse or overuse of the land. Overgrazing by livestock and deforestation for fuelwood are major causes, but the salinization of poorly-managed irrigated land is becoming a very important factor.

The cost of desertification, in terms of lost production alone, is around $26 billion a year. By contrast, the expense of rehabilitating the degraded lands, and halting the spread of deserts, is estimated at $4.5 billion a year. The ratio of loses to costs is therefore 5:1. Even though large sums are being spent, it is not enough and we are losing the battle against desertification.

Why don't governments invest the necessary funds on the grounds that it makes economic common sense? A main reason lies with the status of the people affected. They are "marginalized" people in two senses: they live in marginal lands; they are marginal to the politico-economic structures of their countries. National leaders know they represent no threat to the system if their needs are ignored.

Governments frequently funnel sizeable sums of money into drylands when trouble of a different sort erupts. For instance, in the war between Ethiopia and Somalia in the late 1960s, the superpowers poured vast funds into the zone in the form of military hardware, because the region was so close to tanker lanes from the Gulf. These funds would have been enough to rehabilitate degraded lands in the two countries – and halt desertification along much of the Saharan frontier.

As for the international community, it has not responded as urgently to the problem as the situation demanded. The Sahel disaster of the early 1970s prompted a 1977 UN conference which set the goal of the year 2000 to halt desertification.

THE ENCROACHING DESERT 43

The encroaching desert

North and Central America
Stock-raising imposes a heavy burden on dryland areas.

The world has five major desert zones (hyper-arid lands), which lie in two belts on either side of the Equator. Desertification does not occur in these natural deserts, but in the arid and semi-arid lands.

There are four main causes of desertification each of which is made more acute by excessive human numbers: over-cultivation, deforestation; over-grazing; and poor irrigation. Marginal lands are cleared and ploughed in despairing farming attempts; trees and woody plants are slashed for fuel; livestock over-grazes and flattens vegetation; and improper irrigation sterilizes the earth with salts and alkalis.

South America
Pressure of human numbers allied to rudimentary agriculture are degrading marginal lands. One area in the Atacama Desert has never recorded rain.

Asia
At least 40% of all land is at high risk, and almost one-third is at moderate risk. Much of this is in the USSR, a developed nation that does little to halt the encroaching desert.

Africa
Not only is the Sahel at risk: several other areas are endangered though poor farming practices.

Australia
While little land is at acute risk of desertification, virtually the entire country is threatened by inadequate stock-raising techniques.

Total land area 13.5 billion ha
Risk of desertification
Moderate
High
Very high
Hyper-arid zone

Creeping outwards
Of the Earth's ice-free land area (outer circle), over one-third is already affected or likely to be affected by desertification. The inner core of 6% represents extreme desert. Spreading out from this are the areas at very high, high, and moderate risk – 3%, 12%, and 13% of our land base respectively.

The Sahel disaster
Desertification was brought to the world's attention by the Sahel disaster of the early 1970s. The ostensible cause of the débâcle lay with the drought that overtook the semi-arid zone dwellers of the Sahel. But the main problem lay further in the past. For the previous two decades, the region enjoyed better-than-average rainfall. Consequently, when cash crop cultivation in lands further south (p. 47) started to expand along with the indigenous populations, throngs of pastoralists and cultivators moved north, toward the fringe of the Sahara. Lands that had not been dug for centuries were cultivated, and livestock were crowded on to smaller areas of pasture. (The photograph, right, shows the effects of overgrazing in Niger.) Sadly, these migrant peoples knew, from their tribal traditions, that the moister phase would probably be transitory, which made the drought all the more terrible when it finally struck – more than 100,000 people and 3.5 million head of livestock perished in the early 70s and disaster hit the region again in the early-to-mid 1980s.

The crisis still persists right along the Sahel. Human activities continue to reduce the productivity of lands that speedily degrade under stress.

'The Encroaching Desert' from *The Gaia Atlas of Planet Management*. The use of a narrow range of colours makes it difficult to distinguish between the various degrees of risk of desertification and thereby enhances the image of danger.

word, praised Chinese governmental attitudes towards their environment and people, or, more generally in the atlas, by the criticism of the West and multinationals and the call for redistribution of international wealth. Nevertheless, much of the mapping was novel, certainly in a mass-produced, accessible, paperback atlas of this type. The emphasis was on interdependence, on the manner in which the environment linked often very distant communities, and was, in part, a reflection of such linkages. The first map in the book illustrated the Gaia hypothesis – the proposition that the biosphere operates in an organic fashion using natural feedback mechanisms to sustain life. The atlas contains many arresting photographs and graphs depicting environmental challenges and degradation, and the maps dramatize the same message. In part, this is achieved by concentrating in individual maps on a single topic, such as areas of tropical forest undergoing rapid depletion, or regions of desertification, and ignoring

Opposite 'The Shrinking Forest' from *The Gaia Atlas of Planet Management*. The red zones mark the areas of forest undergoing rapid depletion. The projection emphasizes the Tropics; the absence of names, borders and physical features concentrates attention on the subject; and the background photograph of a wasted forest dramatizes the issue.

any other issue. The use of a world without political boundaries helps ensure this focus.

Graphic symbols are also employed to drive home points. Thus the spread on cash crops, which makes the point that such crops are frequently grown in preference to food for the local population, employs a skull as its symbol for 'country with per capita dietary supply less than 100% of requirements', a vivid way of directing attention to sub-Saharan Africa in particular. More generally, the maps in this atlas 'work' thanks to arresting topics and treatments. They produce a sense of urgency, give issues a spatial dimension and indeed in some cases, as with maps of relative resources, for example per capita water availability, create them by focusing on such a dimension. Julian Burger's *Gaia Atlas of First Peoples* (London, 1990) also stresses the devastating nature of European colonization.

Fragments d'Europe: Atlas de l'Europe médiane et orientale (Paris, 1993), edited by Michel Foucher, contains no fewer than six maps on pollution, including maps on sulphur-dioxide emissions, acid rain and river pollution. The map of the Chernobyl emissions provides vivid evidence of the extent to which environmental disasters can ensure that a widespread area is given unwanted unity. Another map, of pollution in the Elbe basin, again underlines the theme of cross-frontier pollution. Industry in what was East Germany polluted rivers such as the Saale and the Mulde and this was moved, via the Elbe, into the North Sea, producing a concentration of nitrates per litre that is recorded on the map.

The fifth edition of Michael Kidron and Ronald Segal's *State of the World Atlas* (London, 1995), includes maps such as 'Major Chemical Accidents/Incidents reported 1986–92', a map in which the use of red is intended to dramatize the scale of the problem.

This is a world away from the measure of progress by the extension of human control, especially settlement, agriculture, mining and commercial forestry.[19] Similarly, in the aftermath of the 1976–7 Californian drought, the publicly funded and produced *California Water Atlas* (North Highlands, CA, 1978) emphasized the politics and problems of artificial water-delivery systems, and indicated the degree to which they developed out of specific historical circumstances, rather than being natural and immutable. The *Atlas* was offered as a public service designed to present information collected by the government in order to enhance public understanding of the problems.[20] Thus maps were to be used to educate the public; their very appearance expressed public concern.

Similarly, the publication of *The Conservation Atlas of Tropical Forests: Africa* (New York, 1992), reflected concern about the rain forests. Part of a series produced by the World Conservation Union, the atlas set out to map

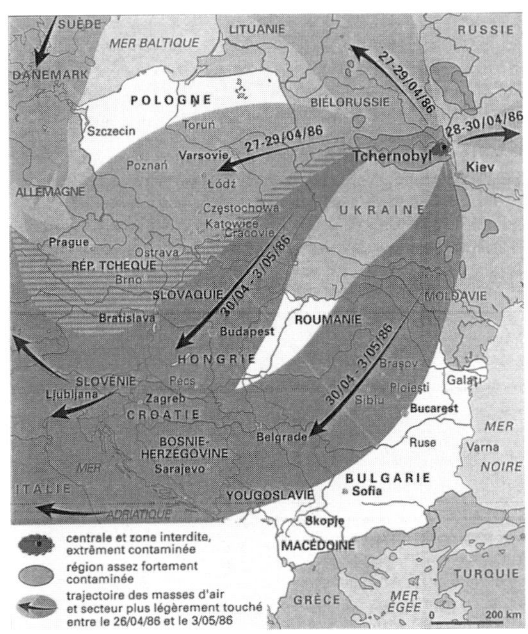

Chernobyl emissions, from *Fragments d'Europe. Atlas de l'Europe médiane et orientale*, edited by Michel Foucher (1993), a bold atlas that sought new ways to map Eastern Europe. Pollution was mapped as an important issue, and was used to show the interdependence of the region.

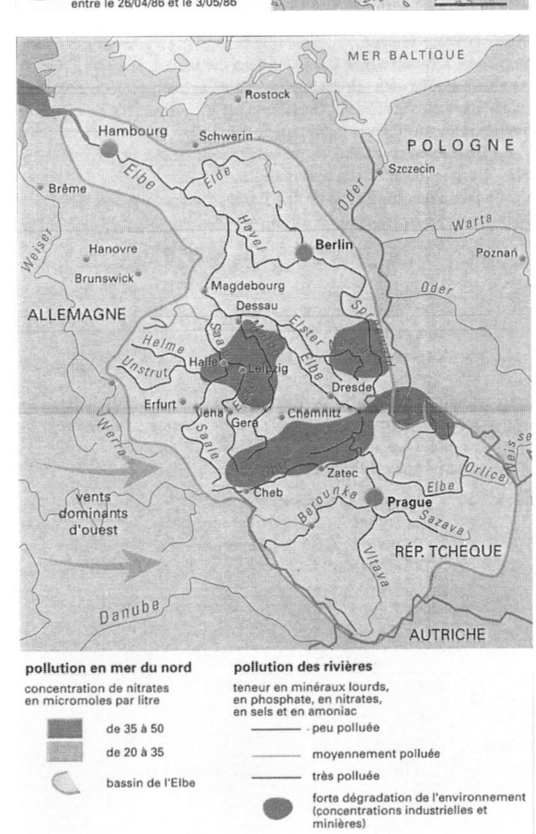

Pollution in the Elbe Basin, from *Fragments d'Europe*. The depiction of rivers in red and yellow helped to arrest and direct the attention of readers. The polluted basin appears lurid and almost pulsing with colour.

closed canopy rain forests and conservation areas. Several of the states had never hitherto had their forests adequately mapped; the mapping made use of new techniques, including satellite information. At a more local level, environmental issues are emphasized in maps and atlases of plant and animal distributions and ecosystems. Thus *Sonoran Desert Plants: An Ecological Atlas* (1996), by Raymond Turner, J. Bowers and Tony Burgess, presented the Sonoran Desert as a fragile and diverse ecosystem.

Controlling resources

Environmental issues readily lend themselves to cartography, as environmentalists use maps in order to highlight risk and their opponents seek to contest this interpretation.[21] Maps can also be used to define and understand bioregions that can serve as the basis for eco-friendly human existence.[22] Yet maps also threaten the environment by revealing its resources.[23] This is not only a current phenomenon. In the nineteenth century, maps were published in order both to reveal resources that could be utilized and to indicate possible supporting transport links. This was particularly the case with gold rushes, as with that to California. In 1851 Charles Drayton Gibbes' *New Map of the Gold Region in California from His Own and Other Recent Surveys and Explorations* was published in New York. Needless to say, the native inhabitants were ignored. In addition, the very process of mapping can be seen as a regimentation of the environment:

> The maps of the West, in their geometric cadastre, set a Euro-American stamp on the land, and indeed allowed this impress to occur. Gone from the map were the flowing lines of stream and vegetation, crossed by the trails of horse riders following the natural rhythms of the land . . . the transition from the birch bark river diagrams of native guides . . . to the crude cadastral diagrams of the Dominion Land Surveyor . . . show the imposition of order.[24]

Mapping areas and sites of environmental interest and concern poses the threat that greater knowledge will lead to more pressure. In its issue of 1 October 1973 (vol. 46, no 1, p. 63) the London-based *Geographical Magazine* published a detailed colour 'Map of Cherished Land', the accompanying text headed 'The cherished land which must be preserved NOW'. This was a map of Britain showing land over which some degree of protection control exists. The purpose of the map was explained as making

> people aware of what exists in the countryside so that they may enjoy it, appreciate it and help to protect it . . . Most aspects of land which must be preserved against the pressures of a changing world are shown, but this is a practical map designed to show the would-be visitor where to go and what to see. It is not a motorist's map but a map for everybody who is concerned with the national

heritage. It is a map urging people to go into the countryside aware of their environment which must be protected for future generations.

Yet this enjoyment also posed problems. Indeed, the map threatened to subvert its very purpose and this led to deliberate omissions: 'some nature reserves are deliberately not publicized to protect particularly rare species or communities', a cartographic 'silence' that offers an interesting corollary to the omissions of defence sites and their layout from maps that some map scholars have discussed. I will maintain the silence by not drawing further attention to the issue, in other words employing my value judgment to conceal cartographic information.

A judgment on the part of the cartographer I found more questionable was the inclusion of motorways but not railways. While it is true that most people reached 'cherished land' by road, it was not necessary to show the motorways, not least because in 1973 they mostly did not cross areas of 'natural beauty', and including them appeared to make a statement that they were crucial to the process, and, more generally, part of the world of environmental quality.

Lastly, it is worth noting that the map adopted an institutional approach to 'cherished land'. It depicted designated areas – national parks, national-park direction areas in Scotland, areas of outstanding natural beauty and proposed areas of outstanding beauty, heritage coasts, forest parks, approved green belts, green belts under consideration, land owned by the National Trust and the National Trust for Scotland, national and local nature reserves, national wildfowl refuges, bird reserves, country parks, forest nature reserves – and designated places, for example field-studies centres, national-park information centres, youth hostels, countryside visitor centres in Scotland, and picnic sites recommended by the Countryside Commission and approved for grant aid.

Such a selection doubtless appears reasonable to most readers, and was obviously a response to the problem of researching the map: acknowledgements were made to the bodies that supplied the information. The selection also appears to deal with the problem of subjective judgments by providing an objective guide. This was of course misleading and, in addition, it implied that to be cherished, landscape had to be validated by authority and institutions. Aside from the degree to which the designation of areas and sites reflected a range of considerations, including funds and opportunities, different policies in particular regions – that, in short, the very objective source of authoritative designation had its own politics, variations and problems as a source – there was the wider question of the values and experiences left out by such an approach. This was a landscape drained of imaginative potency and emotional power by the method of its depiction. More specifically, the finer details of local landscape that were cherished by many were neglected. In Devon, for example, there was nothing between Stoke Woods, a local authority picnic site close to Exeter, and the southern edge of Exmoor National Park. Bar one field-study centre and one forest-recreational area there was nothing in south Somerset. Alas, there was no discussion of such issues in the accompanying text. This was very much a map of 'silences', and of the bureaucratization of leisure and the countryside: experience was there to be zoned. As already suggested, 'silences' do not generally indicate conspiracies; rather they reflect the interaction of problems of mapping – of selection and depiction – and habits of mind, especially convention and convenience. The crucial problem is a lack of self-awareness and reflection, a failure to acknowledge that choices have been made and that the resulting map is necessarily an approximation.

Modern discussion of environmental issues is a crucial part of the problematizing of science and technology. This very problematizing has extended the scope (or 'boundaries') of political controversy, and has helped to ensure that issues of legitimacy now play a central role in science. Maps are important in any discussion of legitimacy, because they can be a way to convey a sense of risk and challenge, and to make it concrete by providing locational specificity, for example by revealing areas of deltaic salination that may arise from upstream water extraction.

Environmental quality is both a resource and an issue. Both are subject to mapping and contention, and mapping is an aspect of controlling the human perception of the environment. More generally, resources in the environment can be staked out through mapping. This interacts with international political contention thanks to the frequency of disputes between states over both environmental regulation and resource control. The former relates largely to the spatial movement of environmental degradation. Thus, air or water-borne pollutants are tracked and maps are used to record and dramatize the findings. This is true, for example, of the movement of pollutants from Alsace and the Ruhr down the Rhine to the Netherlands and the North Sea, or of air-borne sulphur dioxide from Britain to Scandinavia or the USA to eastern Canada.

Resource control is a more immediate issue to governments and companies because it entails disagreement over profit, rather than over cost and compensation. The pressures of demographic and economic growth, the intensification of regional economies and the globalization of the world economy, lead to a situation in which the search for and utilization of resources become ever more widespread and important. The pressures, policies and proposals arising from human interaction with the environment have a crucial spatial dimension and in part can be approached through maps. These developments have been exacerbated by the exhaustion of established resources, such as fish, oil and water, and the possibilities of profitable exploitation in regions hitherto deemed inaccessible or unprofitable.

This position enforces the division of the world surface, for exploration and producing companies have to know from which state they should acquire exploitation and transport rights. Thus the impact of economic demand has been an intensification of frontier disputes, and these have had consequences in terms of contested cartographies. These disputes are not restricted to the land surface, but extend to the sea and also involve issues of underground and underwater rights.

Furthermore, the possibility that resources may be discovered encourages territorial disputes. In 1995, Romania and Ukraine contested the possession of Serpent Island off the Romanian coast as part of a discussion designed to prepare for a treaty between the two states. The Romanians rejected the 1948 protocol under which the island was transferred to the Soviet Union as an imposed settlement, and the Romanian Foreign Minister told the Senate that although the island was not then an asset it might become one, owing to oil and natural gas reserves.[25]

Owing to disputes over resources, the detailed mapping of boundaries is often the cause of controversy and it is difficult to reach agreement. Thus mapping simultaneously exacerbates disagreements and these disagreements make frontiers difficult to map. This is especially so if maps are expected to depict frontiers at sea and thus, also, control over islands. Under the 1982 UN Law of the Sea Convention, islands, as well as mainland possessions, were to be taken into account in defining maritime zones. Islands, such as the Hawar Islands, contested by force between Qatar and Bahrain in the 1980s, and the Red Sea islands, fought over between Eritrea and Yemen in 1995,[26] become contentious as a result of the actual or possible prospect of oil.

Lines on the map are therefore a clear expression of value, and the map can be a device as well as an icon of struggle. In addition, notions of threat and unequal treatment can be presented for economic, as well as the more conventional political and strategic, issues, as in maps devoted to the consequences of dam construction. Furthermore, the map can appear to reify what is often a more complex reality, as in the role of river basins in hydropolitics, and the nature of oil or gas basins. Environmental changes add further complexity to mapping, because rising sea levels and changing coastlines, owing to erosion and sedimentary deposition, alter the delimitation of maritime zones.

Hydro-politics is an area of acute international tension, not least in the Middle East,[27] that lends itself to cartography. States can adopt water-utilization and management policies that affect lower riparian states. Thus India's management of the Ganges harms Bangladesh, and Turkey's damming of rivers is perceived as hostile by Iraq and Syria. Maps of river systems can establish a range of interest different from that suggested by maps of territorial boundaries.

If economic issues interact with international frontiers and frontierization, the same is also true of the domestic sphere, although there issues of control are less clear-cut and their mapping more problematic. An international boundary represents either the resolution or the potential for conflict,[28] but the number of the bodies directly involved is limited and there are clear-cut processes for the delimitation of frontiers and for the allocation of oil and other sub-surface rights. Disputes commonly arise over which or how these should be exercised, but not over the need for a frontier or the question of what a frontier is. Thus the map describes a reality – a division of power – the accurate depiction of which is actively sought.

Consumerism, value and values

This is not the case with the domestic sphere, where the nature of power and the divisions of resources and possibilities that play a crucial role in this power are poorly mapped: there is no clear methodology for the mapping, the relevant map consciousness is only weakly developed, and the agencies of authority do not generally seek to map contention in the domestic sphere or devote cartographic efforts to it comparable with those used for international boundaries. This is despite the fact that the landscape and the cityscape, the areas that have to be mapped, reflect the power to control. At the local level, the landscape, urban and rural, arises from property patterns and these express patterns of spatial identity, inclusion and exclusion. In addition, dynamic links within society shape and are shaped by the presence of marketing and transportation facilities, and these are in part also a product of power, privilege and property. Furthermore, the aims that underlie the use and arrangement of land and city or the creation of infrastructure are contested and contestable.

A polity defines itself, and is defined by others, in part through its cartographic image. The unproblematized nature of the domestic space is part of this image and can be seen as a political statement. Traditionally, the domestic space was presented in terms of incorporating boundary lines depicted in conventional projections and perspectives that actuated the customary image of the state. Thus, Britain or France, as usually understood, was the image, and the domestic space was simply an aspect of it. However, new technology, especially the role of the computer and Geographical Information Systems, makes it possible to re-present space in order to make social characteristics apparent or more apparent, or to give them a clear locational dimension.[29] This is an example of visualization – making visible what cannot be seen.[30] A recent example of this is the use of computer-generated population cartograms to present British census information in order to offer spatially detailed visualization of social information. This offsets the conventional minimization of urban in favour of rural characteristics provided by conventional equal land-area projections; London or New York City as less important than Lincolnshire or Nebraska. Instead, the equal-population maps, with their focus on areas of high population, highlight such shifts as increasing social differentiation in the 1980s.[31]

Maps and atlases reveal social differentiation and also reflect it, although they are presented as scientific, objective products that do not involve such differentiation, and the treatment of subjects by map-makers often minimizes such differentiation. For example, maps and atlases generally ignore agrarian social issues and problems. As, in general, with book ownership and newspaper readership, map and atlas purchasing is socially skewed and preponderantly by the urban and the affluent. Maps and, in particular, expensive atlases are consumer products and those with greatest purchasing power buy more. They are also better able to store, preserve and display their maps and atlases; there are bookcases, and at the appropriate size, for the atlases, and frames for the maps.

A similar social dimension is present in the access to maps stored in electronic information systems, for both the equipment and the information involve costs. The increasing focus of cartographic information and accessibility on electronic geographic information systems will accentuate the social and geographic contrasts present among map-users. This can be presented in an optimistic light, as an inevitable aspect of the different rates at which individuals, groups and peoples move from printed to electronic information, and it can be argued that, in the long term, maps will be accessible to most.

Alternatively, or, in addition, the economic aspects of profitability, cost and income can be seen as establishing a relationship between map-producer and map-user in the field of electronic publishing that limits the latter.[32] To use a term employed by Harley, there will be more 'silences' as far as map-users are concerned and they will be serious. This will be true, both in terms of very varying rates of accessibility between different countries and of great variations within individual states, or, as they might more appropriately be termed, markets.

Such a discriminatory situation will not be novel. Map use grew greatly in the interwar period in both North America and Europe, as car ownership and road use grew. Demand for maps grew and map companies responded actively, producing and distributing readily usable and relatively inexpensive road maps, or maps on which roads and towns organized space.[33] This process can be presented in a progressive light as part of the democratization of cartographic knowledge and map ownership. However, it can also be problematized by pointing out, first, that an emphasis on the needs and interests of road users led to a stress on the automobilization of map contents, so that space was organized and depicted from and for the perspective of drivers, and for them alone. Conversely, the earlier impact of rail transport was such that, from about 1850 until the 1920s, general American maps depicted railways but not roads.[34]

Secondly, car ownership remained beyond the ability of much of the population. This was of limited concern to map publishers: those who could afford cars could afford maps, and those who could not were less likely to purchase maps. Thus the interests of map-users, as mediated by the commercial process, affected the contents of maps, providing a dynamic market that was exploited by producers. More generally, maps are affected by social and legal assumptions and these are linked to consumer preferences. In American maps there is not found the coverage of rights-of-way across private land, public footpaths, that are, by contrast, readily shown in British maps.[35]

Tourism

Affluence and consumerism also play a crucial role in a major branch of cartography: the production of maps for tourists. Such maps vary greatly in purpose and content, but they have one defining characteristic: they are produced in order to reveal an area and make it accessible and convenient to outsiders. The views and interests of the inhabitants are subordinate, if not ignored. This is true both of the 'authentic' experience of travel, represented as being in the secret precincts off the beaten track, where it can be discovered only by the sensitive, true travellers, and also of the vulgar tourist and democratizing and institutionalizing tourism with which the former 'experience' interacts.

The road atlas to Denmark published by Kort-og Matrikelstyrelsen of Copenhagen in 1994 includes a mass of specifically tourist information, such as the location of tourist offices, camp sites, youth hostels, picturesque towns, churches of interest, museums, castles open to the public, gardens, antiquities, other sites, old railways, amusement parks, zoos, viewpoints, golf courses, horserace and motor-racing tracks, yachting harbours and airfields for gliders, and the course of scenic routes. This is useful but, obviously, assumes an aesthetic. The Hans Christian Andersen

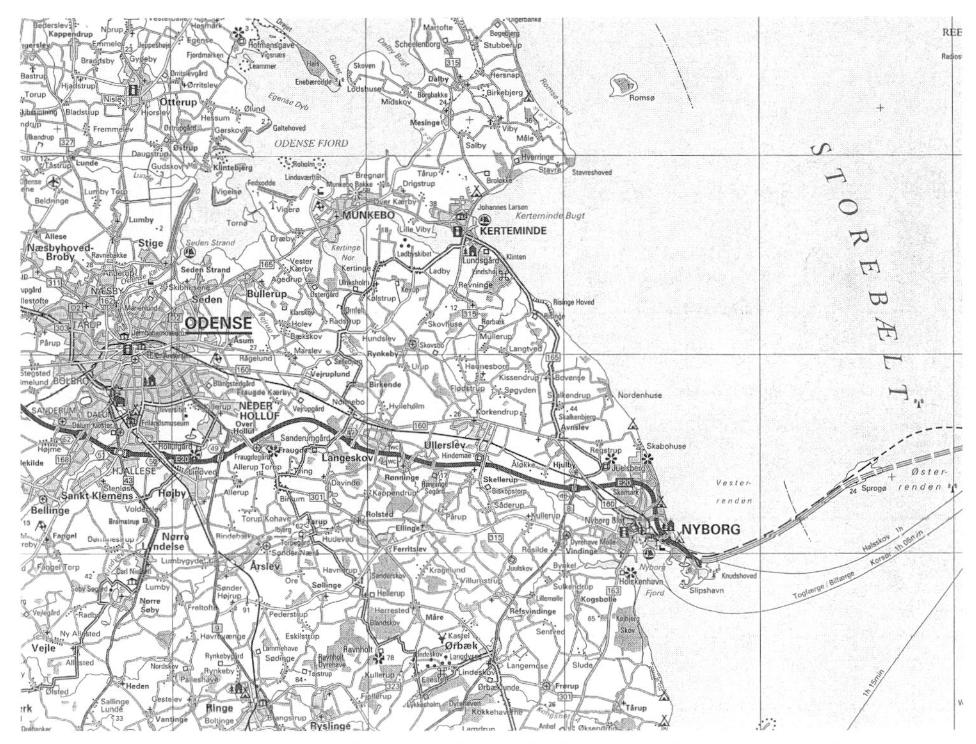

Zeeland, from a Danish road atlas (1994), which represents the country opened for tourism. Numerous settlements are grey blanks, apparently devoid of interest. The use of colour enables the map to convey a mass of information, but this is very much a leisure map. Places of employment do not feature. Even as a leisure map it has limitations; shopping centres do not feature and lavatories are apparently only worth noting if beside the motorway.

house at Odense is marked but not the impressive shipyards. The bridges between Zeeland and Jutland appear as routes; they do not feature as views. Nor does the river route through the low-lying land from Ribe to the sea: the symbols are designed for places not areas; the sole routes that signify are roads.

More generally, in the map, roads link. There is no suggestion that they are problems, not least for the driver, potentially very crowded or fast, as is the E20 across Zeeland. The very variety of symbolization for leisure activity also serves visually to overcome the dullness of much of Jutland, to suggest that if tourists find it dull they are misled. Conversely, there is apparently nothing of interest where most of the population lives. Numerous settlements are grey blanks, devoid of the purple symbols that signify interest, excitement, life. This is understandable in terms of the purposes of the atlas. The 'silences' do not reflect a conspiracy. However, their effect is to further a disjunction between leisure/tourism/interest and the lives and environment of ordinary people than more generally is readily apparent. The sole linkage is the road network.

This dysfunction is true of more than atlases of Denmark. The 1/200,000 (1 centimetre to 2 kilometres) Michelin map of Brittany visually emphasizes roads, but only major routes – bright red passages across the paper. Railways, by contrast, are thin, dark and less prominent. Colour is provided by the red and, less visually prominent, yellow of the roads, and by the green of woods. Towns are grey and make little visual impact. If they contain symbols they are predominantly those for travel – road and rail. There is a symbol for factories, although few are shown. Churches, however, appear in numbers. The map also directs attention to the Michelin Red Guides. Towns that have a plan in these guides are framed in red, while towns mentioned in the guides are underlined in red. As the Red Guides are guides to hotels and restaurants, it is not surprising that the device of frame and underlining both directs attention to the town as leisure space and creates a hierarchy accordingly.

The emphasis on the major roads reflects both the commercial interest of the Michelin company – rail travellers are less likely to be tourists or to purchase maps – and the nature of tourism, in particular, the Janus-faced nature of space as something to be enjoyed or to be overcome. For the former, the map provides scenic routes and details of a great number of roads: the space of the map is repeatedly crossed and thus appropriated for the driver who wishes to see Brittany. For the latter, the tourist who wishes to move as fast as possible to destination, to the *gîte* or camp site, beach or historic town, the map shows how space understood as distance can be mastered as rapidly as possible.

Contemporary conceptions of civilization are reflected in the response to urban and rural society and structures, and this response both replicates and reinforces ideologies and the socio-cultural order. Travel provides an instance, in particular, of the wider metaphor of the city/urbs as a reflection of society; of shifting meanings that urban landscapes accumulate and lose.

These are reflected in maps, in the subjects they depict, the manner of the depiction, and the meanings brought to them by readers. For example, rectilinear street patterns, the grids of American cities, arouse different responses, because such patterns have widely different connotations. The same is true of city squares. To most 'progressive' nineteenth-century European minds a town plan revealing twisting, narrow streets evoked undesirable impressions. Such plans suggested cramped towns, their close-packed medieval centres associated with dirt, disease and poverty.

Yet many modern tourists seek to visit such towns, now cleansed by municipal regulation, commercial pressure and technology. Thus, a jumble of close-packed streets in a map in a Michelin Green Guide will attract the tourist. The response of the latter might be very different, however, if they were considering where to purchase a place to live.

Maps, history and sacred spaces

Thus, aside from directly drawing attention to historic monuments, maps can also suggest the presence of the past in the townscape. More generally, maps act as a form of memory, a reminder of the presence and impact of the past. This is true not only of the obvious reach of geological time and processes into the physical environment of everyday life, but also of such repeated and dense workings of the human environment as place names, settlement patterns and transport routes.

Maps do not simply facilitate and record such processes; they also make them appear natural. With reference to time, they do so by presenting the relationship of past and present in terms of continuity, not conflict. There are, of course, battle sites. Furthermore, nomenclature sometimes indicates past and present dispute; as with the 'Îles Anglo-Normandes' on the Brittany map, not the name the British use for the Channel Isles. However, in general, the map of the present-day landscape no more draws attention to the past conflicts that have played a role in its moulding than it focuses on present disputes. This is true both of 'conflict' between man and nature – deforestation or drainage – and between humans, such as land clearances. Such processes may be implied, especially in detailed maps, but the reader generally has to supply them.

This situation is misleading as a guide to the human (and physical) landscape, but even more to the emergence and nature of politico-cultural-religious spaces. Mapping of the latter is difficult, although such spaces play a potent role in mental maps. To take Europe in 1500–1750, it is clear that the (apparently) precise methodological problems posed by the depiction of international political frontiers provide only a very partial guide to the nature of spaces in the period. The culture that had spanned and united Catholic Europe during the Middle Ages broke down as it was challenged by a number of developments. Some of these, such as the Protestant Reformation, were clearly disruptive, while others, such as the rise of printing in the vernacular, were less obviously subversive. This shift was related to the decline of the universal Christian world view and the rise of scientific accounts that were socially more specific and also to a laicization of culture expressed through the differentiating focus of different languages.

Some of these processes are readily mappable, although the meanings communicated and, in part, created by the maps have to be supplied by the reader. For example the Reformation involved a reformation, or rather reformulation, of religious and ecclesiastical space and spaces. Both the definition of such space and the interaction of structure and process, structures and processes, that space betokens changed. Thus, in Protestant Europe the units and structures of ecclesiastical government were transformed, while the sacred spaces associated with pilgrimages and relics, the lives of saints, holy wells and an entire landscape of providential intervention was swept aside. Sacred space was destroyed. So also was monastic space. Instead, ecclesiastical order and authority became more predictable, less affected by exemptions, by 'peculiars' (areas of other ecclesiastical jurisdictions) and by the impact of international religious bodies, especially the Papacy and monastic orders. Similarly, time was affected, with the abolition of feast days.

Some of these processes can be mapped, but the spiritual landscape that resulted is difficult to capture. So also is the tension between the world of authorized and approved Christianity and religious practices and emphases that were banned or disapproved of, but that, nevertheless, enjoyed public support. In 1765 Richard Browne visited St Patrick's Purgatory, an island in Lough Derg, County Donegal, Ireland, where each year over 10,000 Catholics made a pilgrimage, a practice expressly forbidden under the Popery Act of 1704. He found

> a multitude of both sexes . . . in one place there are built seven small places of a circular form like pounds in which place the penitents are obliged to run so many times round bare foot on sharp pointed rocks repeating so many ave marias etc., in commemoration of the seven deadly sins . . . in other parts they are obliged to wade to the middle in the water and stand there for a stated time repeating a certain number of prayers, when this is over the next penance is to retire to a vault made purposely, where they must remain 24 hours without eating, drinking, speaking or sleeping, for they are sure if they do either the Devil has a power of carrying them away . . . the last ceremony is washing in the lake, when they wash away all their sins.[36]

Sacred space and spaces, understood in terms of a powerful and comprehensive religious awareness; ordered in the minds of participants, but not noted by those who mapped (or today map) the religious life of the period. The same is also true of contemporary treatment of religious life, not least in terms of such issues as enthusiasm, beliefs that create a spiritual landscape on the surface of the earth, and assimilation. The latter is crucial to religious vitality, especially in Europe, where religious groups generally face the challenge of the assimilation of their members into predominantly secular cultures.

Such reflections may seem somewhat distant from tourist maps, but, again, they return us to the problems of mapping and the connotations of cartographic symbolization. To depict the location of churches indicates little about religious life or even the spatial dimensions, in terms of range, intensity and meaning, of the religious experience of individuals, communities and peoples. This also provides scant guidance as to the interaction of past and present. The Brittany map offers another instance of this problem. It shows the prehistoric alignment at Carnac, stone testimony to a past cosmology. Mapped in detail such stone alignments reveal, as at Stonehenge and Avebury in England, a landscape shaped by man that answers to a religious purpose, but the map can only communicate so much. On general maps, the locating of such sites is part of the treatment of the past as consumer product. They take their turn with golf courses.

Conclusions

To locate is to display and thus to advertise. Maps not only play a major role in advertising, as in the frequent use of the shape of Texas – a shape readily understood outside as well as within the state – in order to define the source and style of Texan products and encourage purchaser loyalty.[37] The very contents of maps and atlases also reflect a consumerism that is an aspect of socio-economic differentiation, a situation that is, and was, true both of the domestic and of the international market for such products.[38] The advertisement for *The Economist Atlas of the New Europe* carried in a sister-publication, *The World in 1994* (London, 1993), claimed that the atlas would be not only useful but also 'the ideal gift for a friend or colleague'. Business colleagues did not expect maps of the distribution of illiteracy or of bingo halls. Thus, alongside the analytical problems of mapping socio-economic situations discussed at the outset of the chapter, there is the contextualization of the map as product, product not simply of a series of technical processes but also of the opportunities, needs and operations of a modern capitalist society.

Placing the map in such a context underlines its political character, for the economic process reflects the nature and demands of specific political societies. There are also more direct relations between socio-economic and political dimensions of mapping. In the Western tradition both have been motivated by a desire to explain, classify and organize space. This entails a degree of 'control', in the sense of understanding, but also control insofar as the presentation of space is easier if the space to be mapped is clearly and comprehensively organized, whether in a political or in an economic sense.

This ease of depiction is linked to the issue of appropriateness of use. Before the recent emphasis on environmental issues, there was a strong sense within the West that territory should be developed and that such development entailed political organization on the Western model – bureaucratic, hierarchical, and spatially clear cut – as well as the maximum utilization of the region for economic purposes. This was seen for example in the 1930s Land Utilisation Survey of Britain, in which the information-gathering and depiction process was seen as facilitating planned land use, and planning more generally. Non-agricultural uses of rural regions were presented as non-productive.[39] Similarly, *The Times Atlas of the World* (London, 1968) included a spread on 'World Climate and Food Potential' that offered 'a simple guide to areas of unexploited food-growing potential'. Measures required included drainage, fertilizers, irrigation and mechanization. In short, the environment was to be transformed. *The Reader's Digest Great World Atlas* (2nd edn, London, 1968) was also in favour of applying 'scientific and improved technical methods of farming'.[40]

A particular notion of the value of economic development was given explicit political reference in atlases and maps of Israel, which presented agrarian progress as a result of Zionism, both before and after the foundation of the state of Israel. Zev Vilnay's *New Israel Atlas* (London, 1968), the translation of a Hebrew work, described how 'a parched, water-starved wasteland is being transformed into a fertile, closely settled region'. In the Hula valley malarial swamps were cleared by Jewish settlers. The Jezreel Valley 'in the wake of recurring wars and misgovernment ... became an area of pestilential swamps and was virtually abandoned ... settlers undertook the dangerous and laborious task of draining the swamps and transforming that malaria-infested region into the fertile, fruit-bearing plain it once was'.

Agrarian development was linked to security:

> the population of the Moshavim [agricultural settlements] toiled hard and devotedly, for many years, to turn areas apparently unfit for cultivation into fertile fields. Together with the settlers of the Kibbutzim, they stood fast against harassment and attacks from hostile elements, and they made a vital contribution to the country's security and economic consolidation.

As for the Kibbutz movement,

> it was their determination and their perseverance that turned Palestine's swamps and deserts into a fertile garden; and it was their steadfast courage that helped to decide the issue when Israel's fate was put to the military test.[41]

Arab views were not presented, in map or text, but this was scarcely surprising, for both the maps and the text in Vilnay's atlas asserted the identity of progress and Zionism and the redundancy of those who resisted progress.

In many atlases and maps published in the past ten years, especially in

North America and Europe, some attention has been devoted to a view of the rural world that does not present it as zoned for agriculture, specifically to recreational land use and to the value of undeveloped land. Aside from the degree to which such designation of land use reflects the ability of influential groups within the overwhelmingly urban world of elite opinion to use their influence in order to control the utilization of rural areas, this stress is an aspect of a new appropriateness that arises from a re-examination of notions of progress, a re-examination that encompasses the authority and power of the state as well as the grand project of control over the environment and economic production. Atlases thus partake in a reconceptualization of power. This entails a reconsideration of space, no longer to be seen largely as a field for the implementation of political and economic strategies of control and expansion; and, if as a problem, no longer to be seen in terms of the friction with which distance and geographical factors affected such strategies. Instead, these very strategies are questioned and, as shown in this chapter, the contents of atlases have accordingly changed and become more of an issue.

4 The Problems of Mapping Politics

Problems of analysis and problems of depiction affect the mapping of what is conventionally seen as politics, as they affect the mapping of other subjects. This section was written upon my return from lecturing in Scotland, and the problems of mapping Scottish politics in 1996 were immediately apparent. How is the widespread and deeply felt negative response to Conservative government since 1979 to be mapped? It is more subtle and complex than simply interest in devolution or independence, but, even if attention is limited to the latter, it is far from clear how they should be presented. A map of the constituencies of MPs favouring independence or devolution provides scant guidance to the myriad issues involved, and the contrasts between the views of Scottish Nationalist MPs and those of their Labour counterparts (who themselves are very divided) are as marked and important as those between either and the Conservatives. Devolution has different meanings.

Thus, mapping opposition to the Conservatives provides little guidance to the complexities of four-party politics in Scotland (the Liberal Democrats being the fourth party). There is also an important regional dimension to this four-party world. Labour, for example, is the dominant party in Strathclyde, but makes little impact in north Scotland; there, the Liberal Democrats are more important. Maps can bring out such differences, but their capacity to explain them or capture their nuances is more limited.

Depiction is also a problem. Maps of Scotland are visually dominated by the vast extent and prominent position of the Highlands and Islands, but these lands contain few people. Instead, over 80 per cent of Scotland's population lives in Edinburgh and Glasgow or within 10 miles of a line drawn between them. Any equal-area or real-space map thus exaggerates the role of thinly populated areas *and* conversely minimizes that of the Central Belt. In so doing, it fails not only to heed the importance of the latter, but also the extent to which concern about the views of Strathclyde (the conurbation based on Glasgow), where much of the population lives, influences responses in other regions.

Yet a map proportional to population distribution fails to capture the political experience of regions outside the Central Belt, because the people there do not see themselves as pale echoes of Strathclyde. For them, real space is of political importance and the traditions and views of specific localities are not captured by demographically weighted maps, such as those employing contiguous equal population cartograms that are the basis of Daniel Dorling's impressive *New Social Atlas of Britain* (Chichester, 1995). Scotland scanned from Inverness looks very different from Scotland scanned from Glasgow, and the differences in the panoramas are partly located in the sphere of mental assumptions.

More generally, power and politics are about symbols and issues that are variously open to mapping. A recent atlas of eastern Europe in the twentieth century included a map of Stalinism that indicated the pervasive iconic nature of Stalin's presence, depicting mountain peaks named after Stalin, towns named Stalin or after Stalin, and yet also the multi-faceted spatial nature of Stalinism, with labour camps, major prisons and prison camps.[1] The inclusion of the last item makes an important point about the nature of Stalinist rule and maps it; although it is unclear that the spatial distribution of such sites is of much relevance. However, the map serves to make a statement about the nature of Communism by the inclusion of the topic and a suggestion of quantity, rather than location. Such sites were not marked on maps produced during the Stalinist period, or they were presented under misleading descriptions. Stalinism, like other totalitarian regimes, worked in part by creating an all-pervasive sense of surveillance and fear, a regime that was felt but could not be seen or located: prison camps existed but few knew their location or extent. Terror works on ignorance, on the ungraspable nature and undefined scope of the arbitrary power of the oppressor. The authoritarian state needs to locate its opponents, to understand and control dissidence, but does not itself wish to be circumscribed or understood spatially, other than as a comprehensive and uniform force. The inclusion in the atlas of a map of pollution in eastern Europe, which is accompanied by an explanation that pollution was high under Communism because of its obsession with production, also makes an important statement about a 'silence' that was pervasive during the period of Communist rule.[2] However, the role of the reader is, as ever, important. A map of Stalinism that included labour and prison camps can be seen as a map that indicates the character of Stalinism or this can be extended to characterize Communist rule, a conclusion that may not be intended by the compiler. The reader is left to decide how best to integrate pollution into the overall view of Communist eastern Europe, or indeed, Communism.

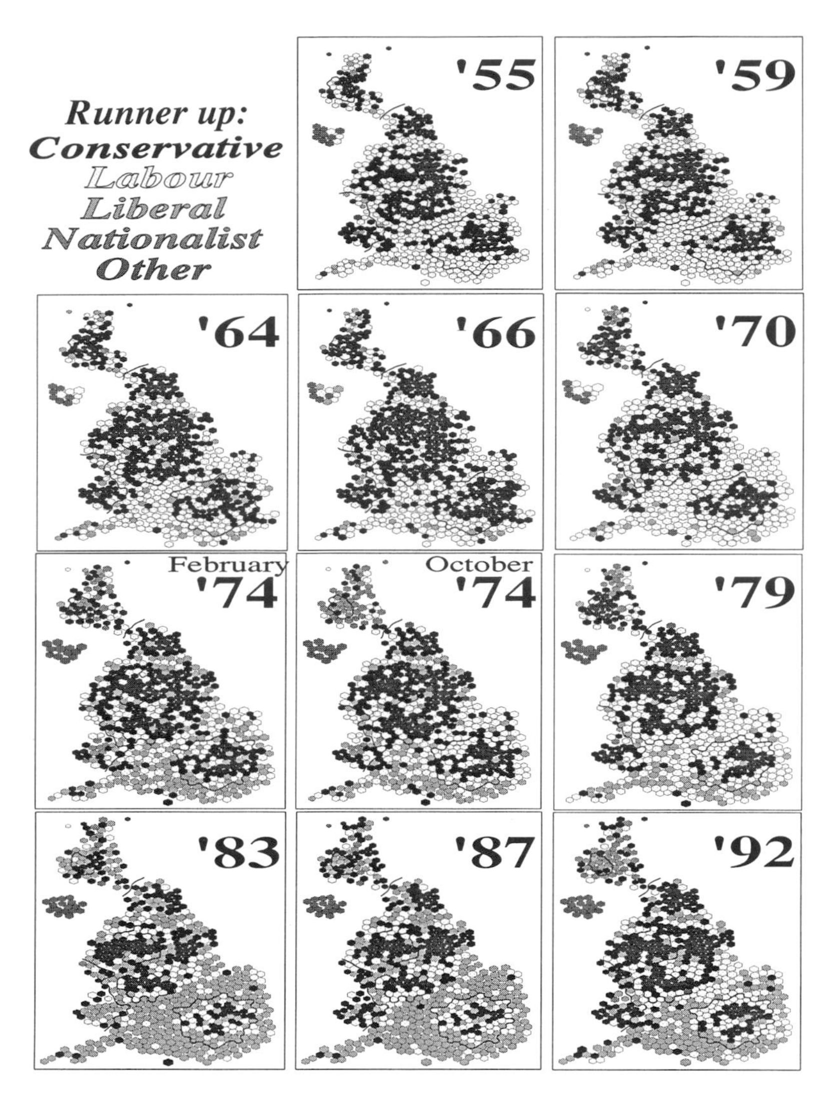

Map of recent British elections employing equal population cartograms, from Daniel Dorling's *New Social Atlas of Britain* (1995). This demographically weighted map is more useful as an electoral guide than as an equal-area map.

Imaginary worlds

Aside from a typology of analysis and depiction, the problems of mapping politics can also be presented in terms of international and domestic politics, the mapping of a system of units, and of the interior and internal dynamics of an individual unit. Each is multifaceted, and present at different scales. Such problems are not restricted to the mapping of the politics

'East European Stalinism' from Richard and Ben Crampton's *Atlas of Eastern Europe in the Twentieth Century* (1996). A reminder of the oppressive nature of power under Stalinist rule that acts as a valuable counter to bland portrayals of territorial control.

of real communities. Mapping imaginary communities and species[3] – Barsetshire to Middle Earth, the world of Long John Silver to that of Winnie the Pooh and Toad of Toad Hall – makes it possible to map in the absence of the politics and polemics of the real world and without the risk of complaint. These worlds have to be created for readers who know nothing of them. Knowledge of the spaces and spatial relationships of these worlds cannot be assumed and therefore they have to be laid out, far more so than in stories located in the 'real' world. Nevertheless, it is still clear that there are major problems with capturing the nature and dynamics of power in such contexts. J.R.R. Tolkien's map of Middle Earth, in his *The Lord of the Rings* (London, 1954–5), a map that was subsequently published separately in large format and in colour, gives no real sense of the spatial range and potency of wisdom and evil, good and ill, that are important themes in his narrative. The ring that is fought for transcends space and cannot be expressed or explained in it. The potential of this ring is expressed in the verse that opens the work:

> . . . One Ring to rule them all, One Ring to find them,
> One Ring to bring them all and in the darkness bind them
> In the Land of Mordor where the Shadows lie.

The power and ambition of the evil Sauron are scarcely captured by placing Mount Doom and the Dark Tower on the map. Mordor is an imaginative realm as much as a place. Instead, at the climax of the book, Sauron's power and powers are fully displayed:

> his Eye piercing all shadows looked across the plain . . . At his summons, wheeling with a rending cry, in a last desperate race there flew, faster than the winds, the Nazgûl, the Ringwraiths, and with a storm of wings they hurtled southwards to Mount Doom.

Map that.

Understandably, Tolkien provides a map of location that clarifies routes, but not their significance; indeed, the text in this sense offers far more explanation than the map. The map of Wilderland in Tolkien's *The Hobbit* (London, 1937) is less problematic because, although also an epic opposing good and evil, this is a book that is more domesticated and less dominated by magic and the occult than *The Lord of the Rings*. Furthermore, the politics of the book is much more expressed in the hazards of the journey depicted on the map. Journeys are central to epics, from Exodus and the *Odyssey* on to such modern epics as *2001: A Space Odyssey*. Movement through space in epics has a spiritual as well as a physical dimension, and, indeed, physical hazards are, in part, to be understood as a spiritual challenge and conflict. Maps are important in the description of the physical journey, but are less useful in its explanation.

To take another, more limited and domestic, level, in which maps cannot fully capture the politics of an imagined world, the pictorial maps printed as endpapers in the HarperCollins edition of Kenneth Graham's *The Wind in the Willows* (London, 1994) provide little sense of the menace of the Wild Wood. Far from being 'low and threatening like a black reef in some still southern sea', as it is described in the text, the wood in the map is an innocuous mass of foliage, not noticeably more sinister than the wood near Toad Hall. Furthermore, the nomenclature used on the map is misleading. There is Toad Hall, Mole End, Rat's House and Otter's House in the landscape, but the Wild Wood appears with Badger's House. There is no mention of its numerous and very different inhabitants, the weasels. Thus, the map creates an impression very different to

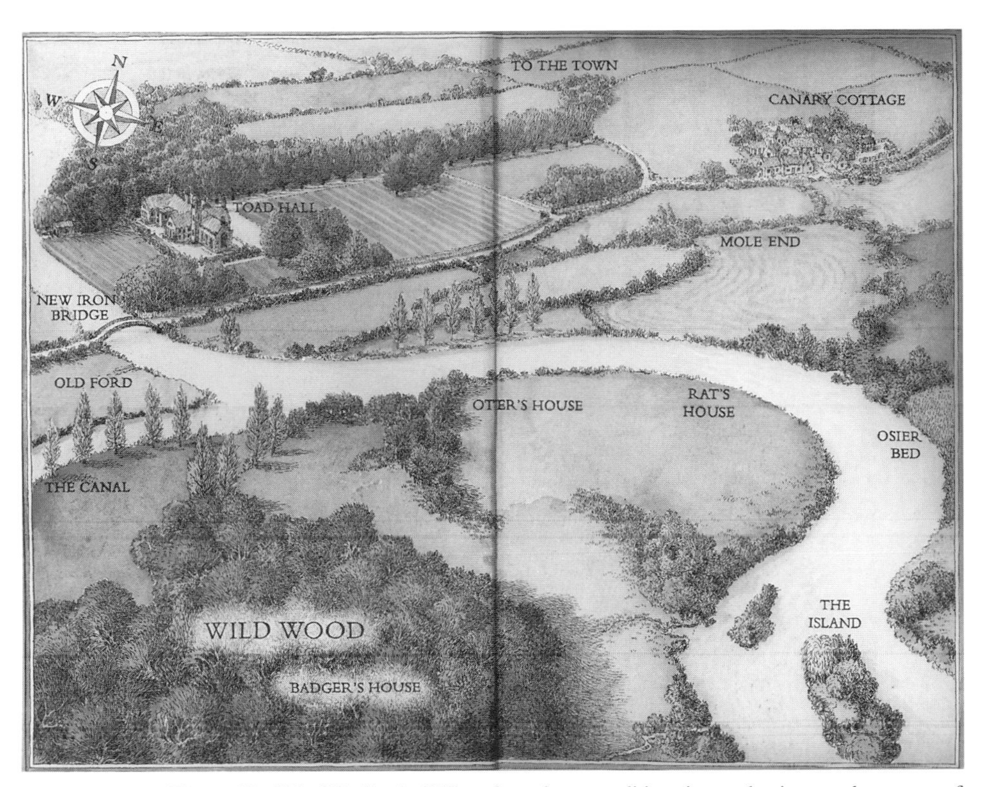

The world of *The Wind in the Willows* from the 1994 edition. A map that ignores the menace of the Wild Wood and acts as an attractive illustration.

the text. The text is a story in which conflict plays a major role, and one in which the Wild Wood and its inhabitants, bar Badger, are the 'other', a challenge to the rest of the non-human world. No such impression is created by the map, a map which, incidentally, is not required in order to elucidate the story. Instead, the map is an attractive illustration, an equivalent to the pictures on the covers: pictures of Rat and Mole, for who would want to use weasels in order to sell a book?

Other imaginary worlds are far more potent. The mapping of religious worlds – of the myths, cosmological understandings and earthly perceptions of different faiths – was and is an exercise in the depiction and projection of power. The resulting cartography created a sense of past and present linked to a common purpose. Religious sites and routes testified to the potency of faiths and to the role of the landscape as a record of divine intervention and thus purpose, a purpose that transcended time.

In the Western tradition, Biblical mapping served to proclaim the accuracy of the Biblical account, especially when confronted by criticism. Such mapping was especially common in the nineteenth century, when it was felt to be important to demonstrate the scientific credentials of the narrative. It was therefore necessary to locate the places mentioned in the Bible to show that they existed even if the location of Heaven and Hell could not be mapped. In Conan Doyle's Sherlock Holmes short story 'The Disappearance of Lady Frances Carfax', Dr Shlessinger, apparently a devout missionary from South America, 'was preparing a map of the Holy Land, with special reference to the Kingdom of the Midianites, upon which he was writing a monograph'. This helped to serve to mislead both Lady Frances and Doctor Watson, for Shlessinger was in fact the villainous Australian Henry Peters.

However, the mapping of the Biblical world offered little help in the understanding of theology. The inter-penetration of worlds central to any providential account poses problems because it destroys the customary intellectual and physical boundaries within which modern cartography is pursued. Similarly, the Augustinian notion of Cities of God and Man that transcend earthly boundaries of time and space does not lend itself to conventional mapping. Yet to neglect such understandings of space is not only to reduce the value of maps of the religious dimension of life, which is commonly presented only in terms of confessional sway – here be Buddhists – but also to underrate the dynamic of the religious consciousness in politics, the sense that divine purpose has to be pursued actively on Earth and through all the spheres of life. Thus Iranian or American fundamentalists have a different spatial sense, but one that, in destroying the boundaries of Earth, Heaven and Hell, greatly affects their understanding of Earthly politics.

Omissions and politics

Omissions are central to the problems of mapping politics, but omissions relate not only to what is not, or possibly cannot be, shown in maps. They also relate to what is shown. Maps generalize (both spatially and by category), abstract, exaggerate, simplify and classify, each of which is misleading.[4] The truth is not only more complex; it is also the very fact of complexity. The general failure of maps to communicate uncertainty is serious, for both analytical and pedagogic reasons. Analysts require the highlighting of lacunae and problems in the data,[5] but map-makers do not tend to adopt such an approach, although there is no reason why they should not. Graphic fuzziness[6] and variations in the intensity of colour can be employed to present and stress uncertainty.

The problems of omissions have become greater in the last quarter of the twentieth century. The spread of powerful mapping software allowed those with little real interest in maps or knowledge about them to produce maps more readily in order to support their own views and they are often unaware of lacunae in, and the problems of depicting, their data. In addition, the cartography of commitment seeks to define and isolate 'problems' in the 'real world', not to discuss problems in mapping. Similarly, the cartography of design ignores such problems. It is a natural consequence of the use of designers to produce or edit atlases and the use of maps in other formats, for example advertisements.

The standard political maps appear objective: the mapping of clear data. The mapping of psephology depicts the distribution of votes, decisions taken in a limited period of time by a finite group of people faced with a limited range of options. Yet, psephological mapping is employed as an analytical tool, for example to try to relate voting preferences and socio-economic structure, and thus it demonstrates a more general truth of thematic mapping that it encodes the spatial distribution of assumptions in a double sense: the assumptions that are mapped and the assumptions of the mapping.

Many of the dangers associated with mapping (most of which boil down to misplaced precision) are equally associated with the processes of categorization and quantification in general. The problems of mapping psephology would be just as true of a psephological study based entirely on statistical tables with no graphics. An interesting example is provided by the misleading impression given by the choropleth mapping of demographic ratios in areas of greatly varying population density, such as political preferences as between the Highlands and Lowlands of Scotland. To some extent, the problem here lies with proportionality rather than with cartography. Thus in a statistical table the same percentage figure 'means' less for a small absolute number than for a large one, although it may *seem* to mean as much.

However, the map presents an extra problem over and above the table of percentages because it includes an element of mimesis. The number 9 is not nine times as large, or nine times as colourful, as the number 1, but in a map such a factor is often present. Even so, this is a weakness not of maps in general, but of one particular kind of map, the choropleth. The problem of misinterpreted proportionality would disappear if political support were expressed by pie-graphs and these can be positioned on maps at the appropriate locations; although, admittedly, other problems, such as the overlap of pie-graphs in contested areas, might still arise.

Elections

Politics is all about assumptions: perceptions of interest and values. There is therefore a tension in its mapping. Results can, and are, mapped –

results, generally, in terms of elections. However, these results, the aggregations, decisions reached by individuals in precisely known locations on particular days, are only a guide to the dynamics of politics, more particularly the reasons why people vote.

Two points immediately emerge. First, an emphasis on the mapping of electoral politics – where be Liberal Democrats? – treats the national stage and, more particularly, that of formal parties, institutionalized politics, as the most important. Popular political activity is ignored, minimized, or treated as an aspect of party politics. This is misleading, but understandable: popular political activity is less easy to analyse and to map.

Secondly, insofar as there is one, the explanation of voting offered in maps is limited. One form of explanation is of course suggested by location: if Labour Party voters in England are concentrated in the north, then Labour is the party of the north. This is an example of locational mapping partially suggesting its own explanation. More generally, party affiliations can be mapped alongside socio-economic differentiation. However, the evidence of local, sectoral and regional differences may in fact point to the reasonably uniform national appeal of a political movement; or regional political differences may be primarily related to factors, such as religion, that are independent of socio-economic differentiation or, if related to it, not in any dependent fashion. This raises questions about mapping political activity rather than social and economic data, ties in with emphases on the autonomy of the political, and indicates the problems both of mapping electoral politics and of understanding the maps. Mapping party affiliations alongside religious differences in order, for example, to show that the German Christian Democratic party is more popular among Catholics than Protestants, raises the question of whether religion primarily signifies other identities and interests, or should at least in part be understood in those terms.

Studies of electoral geography have revealed the role of place, of distinctive milieux that are generally multifactual, rather than being an expression of one particular criterion. Such a multifaceted situation presents cartographic problems in depiction and analysis. Furthermore, historical local attachments, local collective memories, were, and are, not rigid: they respond to developments. However, while they can be difficult to map, electoral trends and geographic clusters can be revealed in order to create a complex mosaic of electoral places.

Mapping electoral politics centres on locations, but this is not without problems, although these tend to be standard cartographic issues, for example the difficulty of adjusting for population distribution. Thus in Tasmania the impression of clear Labour superiority that emerges from several elections owes much to the extent to which areas of Labour support have low population densities and large extent, while Liberals are stronger in areas with high population densities. Such maps can be adjusted for population, but are less helpful for elections in which more than two candidates or parties played a major role, for example the 1992 and, to a lesser extent, 1996 American presidential elections, or parliamentary elections in Britain, not least Scotland. The spread on twentieth-century politics in the *National Atlas of Wales* sought to confront these problems. The notes stated

> The maps summarise party representation in Parliament, but this perspective does not give a full account of the political parties. The relative strengths of the other parties contesting the election are as significant as the number and location of seats gained. Further, exact geographic representation of the election results creates a visual impression heavily biased in favour of the sparsely populated, rural seats ... Hence map e offers a diagrammatic representation of the Welsh constituencies with the physical area of each constituency held constant yet retaining all common boundaries. The relative strengths of each party have been plotted using a common scale. Such a diagrammatic reconstruction gives full visual weight to the South Wales area, and the comparative strength of each party is apparent. The scale is based upon the mean proportion of the poll achieved by each party at the ten general elections 1950–79.

There were four of the last maps, one each for Conservative, Labour, Liberal and for relative majority, the last showing the degree of competitiveness associated with each seat. The notes also warned that the dynamism of politics was not strictly explicable in locational terms:

> Political leaders and political ideas changed political loyalties. These historical elements are a necessary complement to the picture conveyed by the maps and diagrams of the rich political sociology of modern Wales.[7]

Aside from the impact of population density on the visual impression of political support, there are other problems with presenting electoral politics. Changes in the electoral system make it difficult to present shifts through time. The boundaries of constituencies,[8] the franchise, the nomenclature of politics, and the identity of political parties, can all be altered. Nomenclature is a particular problem, as it affects the ability to aggregate results for the purposes of presentation. In countries where party allegiance and organization are weak and there are many shifting groups and independents, such as India, it can be difficult to show results in a consistent fashion. Moreover, the locally proclaimed affiliations of a candidate may differ from his activities at the centre of government, not least in terms of the group he aligns with there. In addition, support for the government or opposition does not necessarily mean support for the governing and opposition parties.

Furthermore, in systems where coalition government is central the

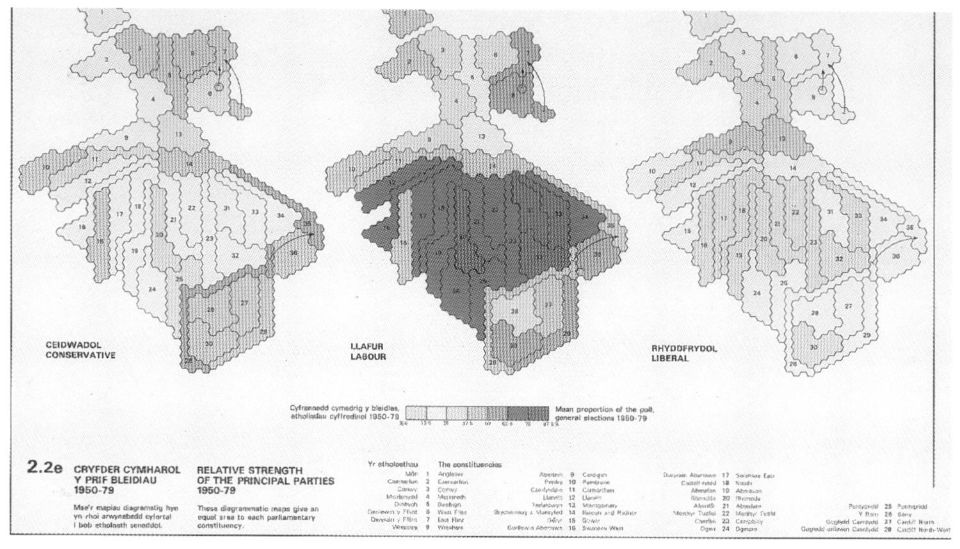

Maps of relative political majorities 1950–79, from Harold Carter's *National Atlas of Wales* (1989). A valuable indication of the competitiveness of each parliamentary seat.

crucial political divides may be different from those suggested by simple divisions such as 'Right' and 'Left'. There can be serious tensions within governing coalitions, as in Germany. Crucial electoral challenges may also be those between parties that would be grouped and mapped together if a dichotomous analysis was to be adopted. Thus in France the struggle between Communists and Socialists has for long played a major role, and President Mitterrand deliberately altered the electoral system in order to increase the challenge to the conventional parties of the 'Right' (themselves divided) posed by the National Front.

Elections have a physical location, the constituency, which can be mapped, but the location is often incidental to the crucial events and nature of a 'general' election. In-depth consideration of voting patterns is rarely provided in atlases, and political campaigning is another generally neglected field.[9]

Elections can also lead to the publication of maps, not for analytical purposes but in order to advance partisan views. The map then serves as an aspect of political propaganda. However, it enjoys more status than caricature, thanks to the resonance of cartography as an objective medium. Thus, in 1884, Rand McNally published for the American Democratic Party a map showing the wide tranches of America given to railway corporations with a pointed accompanying text including the statement 'We believe that the public lands ought . . . to be kept as homesteads for actual settlement.'[10]

Legislative voting can be mapped, but, like elections, it is the product of a complex process that is only partly revealed by mapping. Moreover, there is a dimension to national (and local) politics that cannot be easily treated in this fashion. This is composed of institutional politics and, more specifically, of the politics of representative institutions. The latter are difficult to map, although they have a physical nature that is important. The very configuration of bodies such as the British House of Commons, with their members sitting facing each other on two sides of the chamber, encourage an adversarial system with two sides of more or less internal cohesion. The spatial cohesion and location of the Girondins and the Jacobins within the French National Assembly made their views more prominent during the French Revolution.

By contrast, rooms where there are no sides, where all the representatives sit in straight rows facing in the same direction, for example the Texas Senate, or in a semi-circular fan all looking towards the 'objective' authority of the presiding official, do not reflect or compel such a sense of spatial antagonism or a striving for attention that is based on a two-side dichotomy. The movements of individual delegates within representative assemblies can of course be mapped, by allocating locational values to party affiliations and to rank within parties, but much cannot be mapped. For example, an increase or a decrease in party cohesion can be crucial to the workings of a representative assembly, such as the British Parliament. It can be graphed, but not readily mapped.

Political culture is more difficult to map than political events. Thus, for example, the conceptualization of centre/local relations in terms of regimes of power and knowledge, is difficult to map.[11] The very treatment of knowledge as a form of power and of power as a form of knowledge that has so affected the modern conceptualization of mapping and its contexts makes the mapping of politics difficult because it widens its agenda and sees much of its potency as lying with language and other cultural practices. Political language deals in concepts of identity, inclusion, exclusion and role. They have a spatial dimension in such language, a space that may be geographical – 'We in the north think/need . . .' – or social, as in attempts to define policy with references to social groups and to create such groups in political terms, or may appeal to other possible constituencies. Such rhetoric, however, depends on an absence of cartographic definition, let alone precision. The interest group appealed to or conjured up is generally designed to be as comprehensive as possible, because in a democratic society most politicians wish to emphasize the range and intensity of their support.

Politics can also rest in part on landscapes of fear and anxiety, understandings of location in which there are real or apparent threats. These are not easy to capture in maps, but such perceptions are important. To employ the analogy of crime, crime data is generally incomplete because it offers only limited understanding of geographical and social contexts, including communal views of the dangerous nature of the location of criminal acts[12], in short an often finely understood spatial awareness of zones and boundaries. Awareness of such patterns is also required in the understanding of politics, especially, but not only, of local politics.

Geopolitics

Geopolitics is commonly approached as an aspect of international relations, but there is no reason why it should not also be applied to domestic politics.[13] However, although there was no common definition of 'geopolitik', the ideology of geopolitics was very much one of international competition, as developed most obviously in and by Nazi Germany.[14] Thus, for example, the *Atlas des deutschen Lebensraumes im Mitteleuropa* was published serially in the late 1930s and early 1940s.

Geopolitics rested on the realist theory of international relations, and on the geography of states, as developed by Friedrich Ratzel (1844–1904), and it was presented as an approach to state competition, one in which the territorialization of space was an expression of, and was held in tension by, conflicting political drives. Halford Mackinder (1861–1947) employed cartography in 1904 when he developed his notion of a Eurasian heartland, which was impregnable to attack by sea and threatened to overrun the whole of Eurasia. His was a geography of challenge and threat, and cartography was an illustration of confrontation. Although he disliked the words geopolitics and geopolitician, and wrote extensively on geographical topics, Mackinder is generally considered the leading British geopolitician.

However, both the geography and the shifting technology of international politics created problems for Mackinder. At the chilly meeting of the Royal Geographical Society in London on 25 January 1904, where Mackinder developed his ideas in a paper entitled 'The Geographical Pivot of History', Spencer Wilkinson criticized him for using a Mercator projection and thus exaggerating the size of the British Empire (as well as the Eurasian heartland). Leo Amery, also present, argued that geopolitics and geography would be reconfigured by new technology, namely air power rather than the rail links on which Mackinder focused:

> a great deal of this geographical distribution must lose its importance, and the successful powers will be those who have the greatest industrial basis. It will not matter whether they are in the centre of a continent or on an island; those people who have the industrial power and the power of invention and of science will be able to defeat all others.[15]

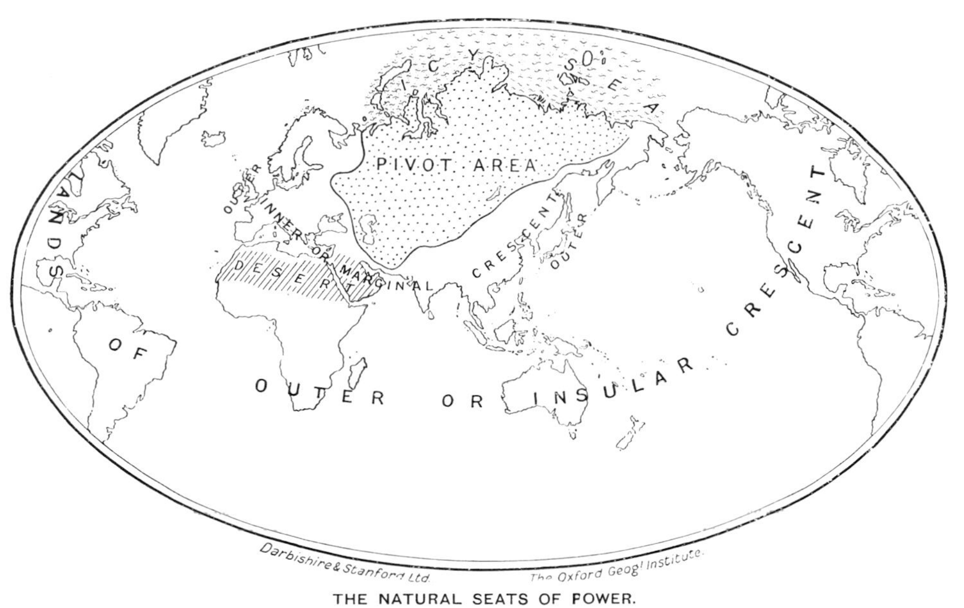

The Eurasian Heartland, from Halford Mackinder's 'The Geographical Pivot of History', *Geographical Journal* (1904). Mackinder argues that in the heart of Eurasia there was the 'geographical pivot of history', control over which would threaten other powers. His survey was a call for vigilance and for a united British empire able to resist whichever of Germany or Russia dominated the 'pivot'.

More generally, geopolitical concepts and mapping did not have to serve the apparent necessity of expansion, or analyse its origin. It was also possible to use geopolitics in order to explain how peace could or should be anchored, as in *Geography of the Peace* (New York, 1944) in which the rimland was presented as dominating the Eurasian heartland. However, whatever the purpose underlying geopolitical maps, there are obvious problems in deciding how best to present state interests, strengths and threats.

Geopolitical mapping was not a constant. It responded to strategic issues and notions, to spatial awareness and to cartographic assumptions. The geopolitics of navalism advanced in works such as *The Influence of Sea Power upon History, 1660–1783* (1890) by the American naval commander and lecturer at the Naval War College, Alfred Thayer Mahan (1840–1914), at the close of the nineteenth century, reflected the immensity of the oceans in the Mercator projection. The 'Cold War' led to a new cartography, which focused on the confrontation between America and its allies and the Soviet world, and made it appear both natural and the best way for the rest of the world to be considered. This approach was expressed in terms of containment and confrontation, themes most vividly illustrated by different perspectives. Thus, confrontation was demonstrated by polar-centric maps that employed a polar azimuthal projection. By being centred on the North Pole, this projection shows shortest distances or trajectories as straight lines through the Pole if a great circle passing through the terminals of the trajectory also passes through the Pole.[16]

This was a suitable image for a world in which inter-continental missiles were apparently a crucial projection of power, for surface geography was less important than shortest distance. Such maps also served to demonstrate the apparent threat to the USA posed by a Soviet Union that was closer over the North Pole than across the Atlantic. Thanks to the shift in mapping, Leningrad (St Petersburg) and New York appeared closer than hitherto.

The development of geopolitics reflected the impact of global strategic issues in the actual and potential clashes between powers from the nineteenth century, and was, more generally, an aspect of the environmental determinism that was so influential from the 1890s. Geography was a central aspect of an intellectual world that linked humans and their physical environment, and maps served to explain these linkages. The political expressions of the impact of space were thus but one part of a greater whole. Furthermore, the very notion of environmental determinism helped to politicize the physical environment and geographical relationships, because these were seen as directly affecting the strength and policies of states.

In domestic political terms, mapping is an aspect of geopolitics for it creates or reflects a sense of the state and nation as a unit, a spatial totality that is central to the map. Furthermore, the process of mapping classifies and delimits units within such a totality, indeed makes regional aspirations spatially concrete and congruent, but generally does so as part of a map in which the political parameters are those of the larger state. The region is thus defined with reference to the state,[17] while the state is seen as the sum of its units, but greater than them. However, the former has historically been the more important relationship and has thus lessened the potential separateness indicated by the depiction of constituent units. Cartography has reduced regions to being parts of a state and has lessened their distinctiveness: indeed, made it simply a matter of constituent status. Maps produce an image of a country/state/nation as more integrated, distinct and centralized than was and is in fact the case. This affected, and was at times part of, the political project of state formation. Alongside defence, the maintenance of frontiers was a major reason for governmental support of mapping. Space was homogenized in order to be reorganized at the dictates of the central government.[18]

In addition, mapping the constituent units of a state provides little information about the practices and processes of power. It is, for example, difficult to show the centralization, or decentralization, of power within a state. Thus, for Britain, in the 1990s, it is possible to indicate county-council boundaries, but far less so to indicate the movement of power from such bodies to central government.

Ideology

Different political cultures dictate, or at least encourage, varying agendas for study and presentation. This is as true of cartography as of other supposedly 'objective' spheres, for example the various branches of scientific knowledge. The role of ideology can be seen at its most blatant in didactic totalitarian regimes, for example Nazi Germany. The cartography may be of a high standard, in terms, for example, of clarity of line, choice of colour or other technical issues, as in Communist East Germany; but the thrust of the map-making, as evinced in the selection of topics for depiction, the accompanying maps and the symbolization employed, is clearly partisan.

Thus the first map of the historical section of the *National Atlas of Mongolia*, a Communist work published in Mongolian in Moscow in 1990, was devoted to the 'Victory of the Mongolian Peoples Revolution' in 1921–3, with accompanying inset maps devoted to the organization of revolutionary societies in Urga, the 'liberation' of Khlagg and the 'liberation' of Urga. There was no suggestion that the earlier history of Mongolia might be relevant. Under the Dzhungarians in the late seventeenth and early eighteenth centuries, the Mongols had been a formidable and widely spread force, and this was even more true of the medieval empire created by Genghis Khan, but such a cartography would reveal a very different relationship between the Mongols and the rulers of Moscow than that envisaged in 1990. Instead, the atlas continued by mapping other episodes in which modern Mongolia had followed the Soviet lead, with maps of the 'Victory over Japanese Militarists 1939' and the 'Participation of the Mongolian People's Republic in the Crushing of Militarist Japan 1945'.[19]

Yet ideology was and is also clearly present in other political cultures and their cartography. This point, forcefully made by Harley and Wood, in their critique of Western mapping, more specifically, respectively, those of Western imperialism and modern American liberal capitalism, can be extended by considering the very notion of global mapping. To produce a map of the world was and is to offer a statement about the relationship between a people or state and the wider world. This might relate to commercial interest, imperial ambition, geopolitical concern or ethnic consciousness, among other factors. Thus, in America, university and school geography and popular cartography took a greater interest in the world, and in America's interaction with it, after the Spanish–American War (1898), and World Wars One and Two, as American intervention in the rest of the world grew, for imperial, commercial and geopolitical reasons.[20]

Nevertheless, direct political control over mapping in 'First World' countries is and was less than in their totalitarian counterparts. Political, rather than commercial, supervision plays a major role in the latter. For example, during the Soviet period, the drafts of maps for the *Armenian Atlas* and the *Armenian Encyclopedia* had to be sent to Moscow for final approval by the Soviet Academy of Sciences, a politicized body, before they could be printed.[21] The Cyrillic alphabet was used throughout the Soviet Union, except in Estonia and Latvia.[22]

More generally, national atlases have an iconic role,[23] either by diffusing a sense of the naturalness of the nation or, more directly, by linking it to a particular political ideology. In the age of nationalism the nation-state was the prime ideology affecting cartography. Thus, for example, the Finnish Geographical Society published an atlas of Finland in 1899, and, in an extended edition, in 1910 while Finland was still ruled by Russia. It was a clear expression of national identity.

National conceptions of space also affected the distribution of cartographic attention, encouraging a concentration on the mapping of particular areas in atlases or map series that purported to be global. Thus, the Rand McNally *New Household Atlas of the World* (Chicago, 1885) devoted four-fifths of its space to the USA.[24] Such an emphasis was common and, although less marked, is still the case with most atlases of the world. The emphasis is a slanted account of the respective importance of parts of the world. It reflects suppositions that may be less bluntly stated than those generally associated with ideological mapping, but ones that, nevertheless, invite criticism. Yet, as already suggested, this is a matter of a spectrum or a continuum, not of black and white, or, to use a less contentious terminology, good and evil. Furthermore, it could be argued that it is psychologically rather than ideologically significant for cartographers, first, to use larger scales for countries that are near them and their readers, and, secondly, to put these countries nearer the beginning of the atlas. It might be asked whether this is any more ideological than for historians to give more space to recent than to remote periods.

Rather than endlessly focusing on problems, it is also important to draw attention to the possibilities presented by maps and mapping. For example, A.J. Christopher's *Atlas of Apartheid* (London, 1993) provided a vivid and novel account of apartheid by focusing on its spatiality and showing it as both important and mappable. This was presented, valuably, at three levels: petty apartheid, urban segregation and grand apartheid. The first entailed detailed social segregation, at the level of parts of buildings and transport facilities, the second a re-spatialization of cities – segregation and movement – as Africans were moved, for residential, although often not employment, purposes, out of town centres. The third was a matter of redefinition and resettlement, as homelands were created by dictat. Similarly, understanding of the Holocaust increases if its spatiality, at both local and European levels, is understood, a theme outlined in a number of recent works, including Andrew Charlesworth's 'Towards a Geography of the Shoah' (*Journal of Historical Geography*, 18 (1992), pp. 464–9) and his 'Contesting places of memory: the case of Auschwitz' (*Environment and Planning D: Society and Space*, 12 (1994), pp. 579–93). At each level of spatiality, for this and other subjects, the role of the state as a focus of ideology and power, and the role of space as an aspect and means of both, emerges clearly from maps.

Sanitized space

The state is a geopolitical bloc in most maps, undifferentiated by any features that relate to politics, other than possibly those of subordinate units such as counties or provinces, presented simply as areas of territory. The state may be described in physical terms or through reference to the mapping of towns or communication routes, but there is no attempt to depict those regions within a state that wield power, or rather from which power is wielded, and those that suffer or are oppressed.

This depoliticization of space links mapping for adults and for children. The latter is politically bland. Thus, for example, the maps in Jane Elliott and Colin King's *Usborne Children's Encyclopedia* (London, 1986) use pictorial images to create an entrancing impression. The map for Asia depicts a Malaysian village for Malaysia and bucolic scenes of peasants planting rice for China. There is no sign of industrialization. Africa is similarly elysian, with an emphasis on animals.[25] In Brian Williams' *Kingfisher First Encyclopedia* (London, 1994) animals dominate the map of Africa. Their competition is natural: a lioness chases a zebra across southern Zaire, while further north a leopard pursues an antelope – although in each case there is no capture and no blood is shed. No human poachers are shown. Lagos, Tunis and Cape Town are pictorially represented, but as attractive; there are no shanty towns. Similarly, the map of Asia is shown as dominated by animals.[26]

Such sanitized space may appear reasonable in the case of children. Few would wish to present them with images of horror, war, poverty and disease. Yet even if only positive images are to be employed, the pictures used

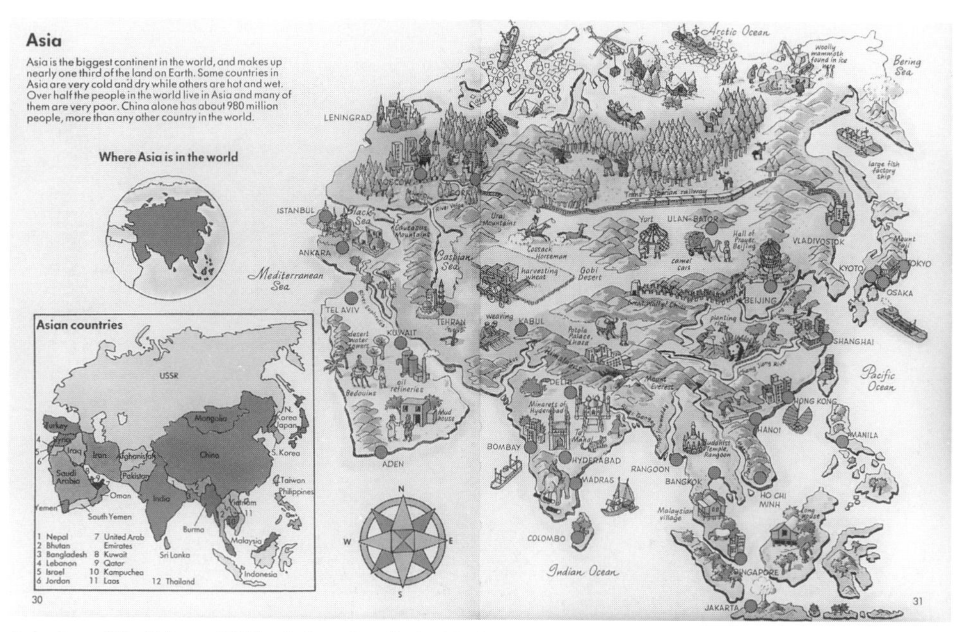

Asia, from *The Usborne Children's Encyclopedia* (1986). A map that emphasizes differences between Asia and Europe and provides attractive rural images for Asia.

are misleading. The stress on animals and on quaint indigenous rituals and activities makes abroad a comforting and unproblematic account of foreignness. Instead, Malaysia or China could as readily be shown using cars and modern buildings, creating a sense that abroad is comprehensible in terms of modernity.

Maps for children create impressions that are generally not discussed in any accompanying text. There is no wish to raise issues that might confuse. To take the coloured two-page political map of much of Asia in the *Hamlyn Children's Encyclopedia* (London, 1992), the same colour is used for North and South Korea and no frontier between the two is shown. Taiwan is coloured in a different shade of green to China; there is no use of a very different colour as with all of China's other neighbours bar India. The West Bank is shown as part of Jordan. The Soviet Union is presented as the Commonwealth of Independent States, with boundary lines between the successor republics but the same colour used for each. China, in contrast, is homogeneous: Tibet and Xinkiang do not feature.

The adult equivalent of the sanitization of space is the undifferentiated bloc of the state, its homogeneous colour unrelieved by zebra or rice-cultivation. Empirically, this is obviously misleading other than as a descriptive account of agreed territorial sovereignty. There is no suggestion of the dynamism that underlies power relationships. Even as an

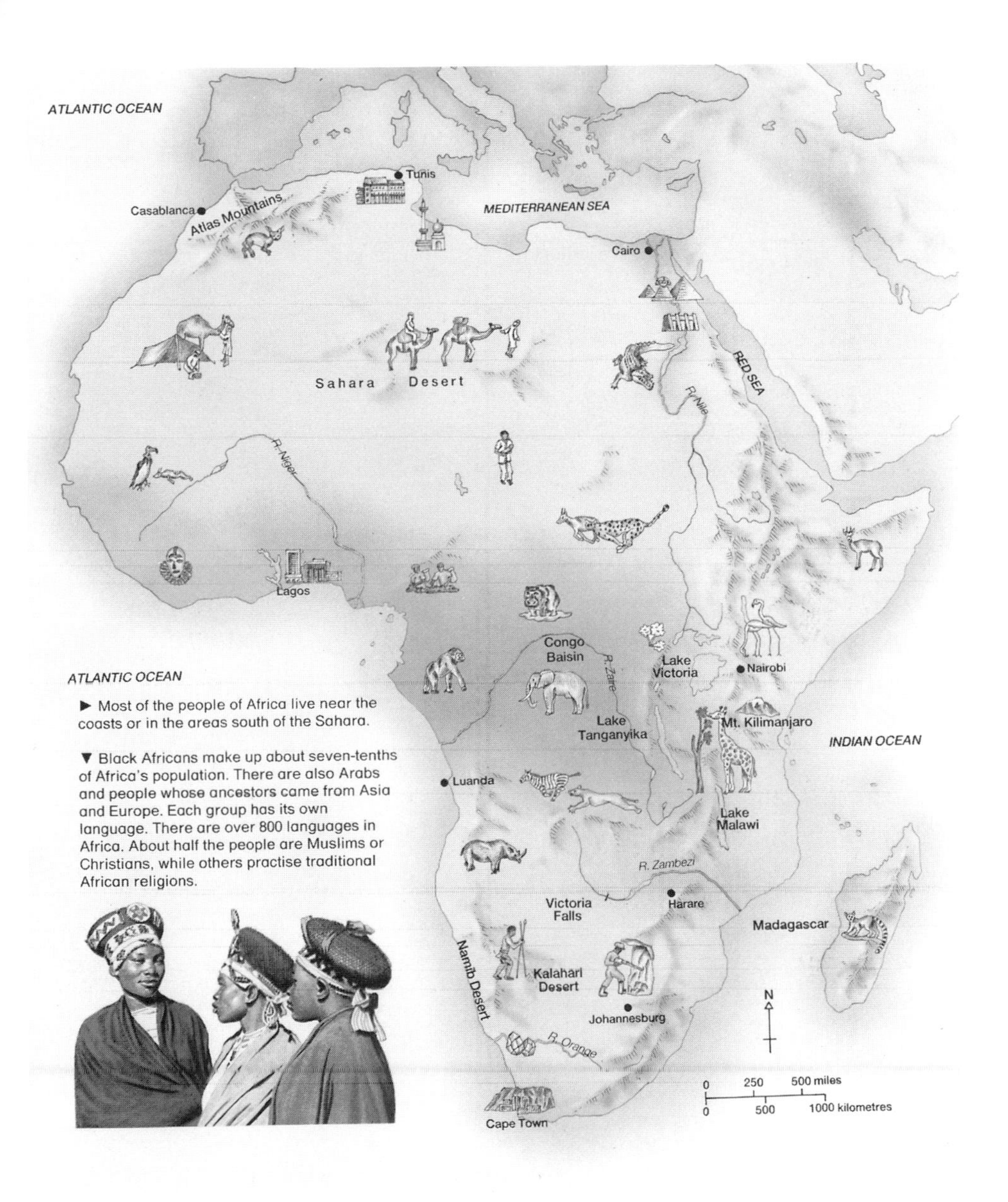

Africa, from Brian Williams' *Kingfisher First Encyclopedia* (1994). Animals dominate this map: they chase but do not kill each other. There is nothing disconcerting either in the natural or in the human world.

account of sovereignty such maps can be misleading. A map of Iraq in 1995 that failed to note Kurdish autonomy in the north, or one of Sri Lanka in 1996 that contained no hint of Tamil activity on the Jaffna peninsula, would be inaccurate. The same is true historically. As the leading history of Iran notes:

> Khurasan produced centrifugal tendencies more than once in the course of Persian history owing to its remoteness from the central areas of Iran and to the narrow corridor that was its sole link with them between the Alburz mountains and the Kavir desert.[27]

Yet such a situation is not depicted or revealed on the maps. Again, the attitudes of mind that underlie territorial control and aspirations are crucial, a point made by Harley, and yet these tend not to be revealed by maps: indeed, the continuity of cartographic idioms can give a misleading sense of similarity. This is seen, for example, in Russian expansion into Siberia and nearby areas. In the nineteenth century this expansion was depicted as a completion of a process begun in the sixteenth century, but

> this expansion must be seen in the more general context of late-nineteenth century European imperialism, whereby a spectrum of factors were operative that were essentially new to Russia ... Whereas the initial occupation of Siberia had to a significant extent been a spontaneous and unplanned process, expansion in the nineteenth century ... was thoroughly premeditated, and had moreover an explicit political dimension ... As part of this, ideological factors such as nationalism, messianism, and reformism, largely absent in the earlier period, played an important role in the nineteenth century. The physical contiguity of the Russian empire should not conceal the historical and geographical breaks in its process of formation.[28]

Again, such a situation is not revealed in the map with its depiction of the organic development of states through the accretion of new territories. Another aspect of the sanitization of space is the habit of not, or only occasionally, mapping domestic political divisions or issues, especially if they are contentious, for example separatism. These lacunae can also be seen in some historical atlases, for example William Pool's *Historical Atlas of Texas* (Austin, 1975), and in many atlases of the modern world. By contrast, a cartographic emphasis on, for example, linguistic or religious divisions within a state can subvert any appearance of space as uncontentious.

Yet, even if power is not uniform, if, for example, the Indian government has only limited control of the section of Kashmir it rules, the territorial claims reflected in historic and modern maps are, nevertheless, themselves indicative of an important reality of pretension. Canada is far from alone in becoming 'a state by being initially ruled on paper'.[29] Maps gave and give regions additional political meaning by allocating them to larger units or linking them with other regions and gave them shape by defining their limits. The very process of mapping areas as part of a larger whole reflects the role of national agencies, such as the US Geological Survey, in creating and sustaining projects of national mapping. Such projects stopped at or close to frontiers.

More generally, maps and their use shaped the geopolitical imagination.[30] They are an aspect of the processes by which the significance of territory has changed over time, processes in which changes in political understanding have played a major role.[31]

Conclusions

'Politics' does not exist in a vacuum. For political ideas, movements and policies to have an impact, to shape the attitudes and responses of people, whether domestically and internationally, they have to be perceived. Graphic images are one of the easiest ways to grasp and influence popular attention, a point appreciated by propagandists. As a result, newspapers and other news media frequently contain pictures, photographs, figures and maps.[32] However, the mapping of news is not free from serious difficulties even when the news in question is of a single event that occurred at a specific place, such as the 1983 Soviet shooting down of a Korean aeroplane. Different impressions were created by the various perspectives and projections employed to depict the flight of the plane.[33]

In its use of maps, the press faces the problems of analysis and of depiction referred to earlier, although maps also offer an opportunity for newspapers. The issue of the role of the state is sometimes referred to directly in the press. An article on global crime in the *Sunday Times* of 17 November 1996 noted that

> Today's criminal sees a national border as an opportunity to exploit and use the laws of one country to try to subvert or do business in another. 'International boundaries mean nothing to criminals,' said Det. Chief Inspector Simon Goddard . . .

The accompanying map indeed showed no boundaries, thus dramatizing the threat of illicitly smuggled goods by ensuring that the only lines on the map were those of smuggled goods: no inhibiting or opposing boundary lines appeared. The map itself served as an example of much newspaper cartography. It was both lurid and misleading. The concentration on Britain ignored the greater complexity of drug and other networks and exaggerated the importance of Britain as a market. This concentration combined with the effect created by the repeated red arrows striking home on Britain.[34] As an example of cartography as drama, the map, however, served the newspaper's point by making themes of danger and threat readily apparent. The map therefore directed attention to the article rather than providing additional information or visual clarification. Instead, the map acted as a form of advertisement for the article, catching the eye of readers saturated with the visual richness of advertisements and the constant theme of crisis in the articles in the rest of the paper.

In confronting the difficulties of assessing how best to move beyond the state when trying to map domestic and international developments, newspapers face an aspect of the more widespread political and intellectual problem presented by the role of the state at the close of the twentieth century. Problems of cartographic analysis and depiction are therefore probed in shifting intellectual contexts. One such, of direct relevance to this book, is the development of 'critical geopolitics'. This seeks to reconceptualize geopolitics away from 'state-centric reasoning, and the privileging of a Western, masculinized, seeing subject as the authoritative, transcendent reader, and practitioner of international politics', and, instead, to investigate points of contact between the geography of international relations and, for example, social theory and feminism. Such an approach, which includes the application of 'post-structuralist' ideas, leads to serious critiques of more conventional geopolitical writing, which can be presented as contingent, not objective, and of the spatial impact and implications of power.[35] Maps can serve as tools of debate, highlighting these spatial implications and thus apparently providing graphic evidence of the nature of the practice of power and of what can be seen as a need to challenge it. There is no reason why cartography has to serve as an adjunct of power.

5 Frontiers

The mapping of frontiers is a central issue in political cartography. In one light this is simply an aspect of a common problem, that of the nature and representation of divides and boundaries in cartography, and, more specifically, the role of lines. However, the problem of lines as a mark and form of division takes on extra weight in the case of international relations, because the lines betoken frontiers and these frontiers are the cause, course and consequence of conflict.

Frontiers also represent an intersection of two aspects of political mapping: domestic and international, although as far as frontiers are concerned the international sphere is far more important. Much of the effort devoted to mapping at the international scale has centred on the mapping of frontiers, and frontier disputes are the single most contentious issue involving maps. Indeed, for many, maps and politics would be essentially an issue of international frontiers, although such an assumption is no longer tenable.

Mapping frontiers is difficult because it requires the consent of at least two parties. Aside from differences of opinion over the course of the frontier, or over where a to-be-determined frontier ought to go, there can also be problems with different mapping cultures, not simply technically, but also involving, more generally, contrasting notions about the meaning of maps. Furthermore, understanding of frontiers is culturally contingent, although the spread of dominant Western notions is an aspect of greater Western control, or, at least, influence, in the world in the nineteenth and twentieth centuries. Though presented as a source, or product of security, of political and cartographic order, these notions, however, have, in particular contexts, been a cause of dissension and strife.

Mapping frontiers is about mechanisms as well as consent. To agree that a frontier should, for example, follow a watershed or a riverline, is not to define it on the ground, nor to chart it on the map. The two interrelated processes both entail considerable problems and each can cause dissension. These difficulties affect the construction (or at least cartographic creation) of 'new' frontiers, but, in addition, are not banished once maps have been drawn. Just as maps are not apolitical, so also are they not fixed. Both in general terms and in particulars, the notion of territorial space and its frontierization are not fixed. Furthermore, the detailed course of a frontier may be a cause of fresh contention. This can be because issues that have been agreed between the two parties are affected by changes in the parties or in their interaction. In addition, changes may occur, for example shifts in watercourses, that require, or may seem to require, a re-examination of the frontier.

Maps are both a means by which frontier disputes are pursued and a measure of them. The appearance of maps in newspapers and on television reflects their role as locators and explainers of frontier disputes. It is through maps that the location and extent of territory in dispute becomes apparent. Maps therefore make frontier differences and disputes matters of note. They are a medium, not a message. Maps are part of the process by which frontier differences are defined, but this definition can lead to an obvious politicization rather than to any peaceful definition.

Maps were used in the recording of frontiers in antiquity, with the Egyptian map of Nubian gold mines, Ptolemaic maps and Roman cadastral maps. Maps were used for European frontiers since at least the fifteenth century. One map was drawn to show a small section of the frontier between France and Burgundy in 1460; another to show the frontiers of the Kingdom of Naples in the late fifteenth or early sixteenth century.[1]

Early Modern European frontiers

Maps were increasingly used for European frontiers from the sixteenth century. A map was used for the negotiations that led to the Anglo-French Treaty of Ardres of 1546, although the negotiations were not without serious difficulties. Before the treaty William, Lord Paget, and a French emissary went with several guides to examine the source of a proposed boundary stream and fell into a serious dispute over which of five springs was the source of the river.[2]

Maps were consulted more often in the eighteenth century in relation to political discussions and in diplomatic negotiations, and were on the whole geographically more accurate. This did not, however, make negotiations easier unless the two powers were prepared to compromise. Accurate maps did not help if negotiators disagreed over the terms that described frontiers in treaties. Such maps did not necessarily help the definition of borders in detail. As in the twentieth century, improved mapping could help to reveal differences of opinion.

On the other hand, if rulers were seeking compromise, improved mapping could help to cement agreement. Such an attitude of conciliation and rational arrangement characterized French policy between 1748 and 1789.

In addition, if it is argued that conflict arose as much from accident and from the opportunism latent in an international system that was imprecise in a number of spheres – most clearly frontiers – as from deliberate planning for war, then increased precision in the mapping of frontiers was as important as the related consolidation of territorial sovereignty and increasing state monopolization of organized violence. All were different facets of the consolidation and spread of governmental authority, and the erosion of the distinctive features of border zones. The implementation of firm frontiers was bound up with the existence of more assertive states and growing state bureaucracies, which sought to know where exactly they could impose their demands for resources and where they needed to create their first line of defence. As a result, the period from the late Middle Ages witnessed a burgeoning emphasis on frontiers throughout western Europe, despite the constraints produced by the limited extent of contemporary mapping. This was not a process restricted to western Europe. For example, with the Treaty of Pähkinälinna between Sweden and Novgorod in 1323 the Swedo-Russian border was documented for the first time.

Improved mapping helped to make the understanding of frontiers in linear terms, rather than as zones, easier, and thus played a major role in frontier negotiations, not least in the attempt to produce more 'rational' frontiers by removing enclaves. This was a hesitant process. Henri Arnault de Zwolle, councillor of Philip the Good of Burgundy, produced in 1444 a map of a contested region between France and Burgundy. This was seen by Duke Philip as part of a process by which French enclaves could be defined and eliminated in order to simplify the frontier.[3]

Nevertheless, frontier rationalization was a hesitant process in western Europe. The notion that Louis XIV of France (1643–1715) wished to create a rationalized frontier between France and the Spanish Netherlands (essentially, modern Belgium) – in other words, to use his power and the capacity for informed reason presented by maps to secure peace – has been queried.[4] Instead, it appears that Louis wished to use traditional feudal claims to push forward the zone of French power. Far from seeking the stability of a clear line, Louis wanted to leave the frontier amorphous and his pretensions ambiguous – i.e. unmapped, if mapping is understood as a form of, and statement for, precision – so that his power could be advanced as opportunities offered.

Even if the impact of Louis XIV's activities was a clarification of the frontier, as Louis' claims were contested and resisted, his expansionism was opportunistic. Enclaves were retained where it suited French interests. Mülhausen remained as an ally of the Swiss Confederation, an enclave in French Alsace, until the 1790s, because the Bourbons did not wish to upset their relations with the Swiss. French fortresses were established to the east of the Rhine and Alps, so that any notion of the role of fortifications in helping to define frontiers has to be qualified by a realization that they could also lend force to enclaves, and thus sustain the zone-like nature of some frontiers. Louis XV wished to conserve French enclaves, for example the towns of Beaumont and Gimay in the Austrian Netherlands. Puysieulx, the French foreign minister, complained in 1748 that the cession in Louis XIV's later years of Pinerolo (1696), Fenestrelle (1713) and Exilles (1713), all east of the Alpine watershed, closed the doors of Italy to France.[5] The lengthy negotiations between France and the Prince-Bishops of Liège over the frontier territory of Bouillon between 1697 and the mid-eighteenth century were scarcely characterized by a spirit of 'reason' inso-far as the latter term is now understood:

> Dans les multiples 'factums' qui vont s'échanger jusqu'au milieu du XVIIIe siècle, les deux parties traiteront en long et en large cette question de Bouillon, en remontrant souvent jusqu'au XI siècle. Elles invoqueront en général quantité d'autorités que nous jugerions aujourd'hui fort sujettes à caution: récits plus ou moins légendaires ou merveilleux du haut Moyen Age, témoignages d'auteurs anciens qui n'écrivent que sur oui-dire, généalogies plutôt fantaisistes, etc.[6]

Alongside the persistence of traditional attitudes, there was also change. Despite the general and traditional conception of sovereignty in jurisdictional, rather than territorial, terms in seventeenth-century Europe, the idea of natural frontiers – apparently readily grasped geographical entities – principally mountains, had become a 'widespread dictum of geographical discourse', readily employed in diplomatic discussion. In the 1659–60 negotiations over their new frontier in the eastern Pyrenees, both French and Spanish negotiators advanced geographical claims as well as historical arguments, although the eventual frontiers 'were neither historical nor geographical but rather a compromise resulting from a bitter diplomatic struggle'. The frontier continued to give rise to negotiations and disputes, a process in which territorial disagreements interacted with the very process of territorialization in which mapping played such a role. In 1688, the Spaniards proposed that a section of the border at Aldudes in the western Pyrenees, currently undivided and common to both nations, should be partitioned. Negotiations over the Pyrenean frontier, however, were still encountering serious difficulties in 1775, and the Caro-Ornano Commission of 1785 failed to settle the matter. The French Revolution led to a new departure, transforming the meaning of lines on a map, with the 'politicization of natural boundaries', and the instilling of a stronger and more firmly policed sense of national territoriality in border regions.[7]

The selection, definition and mapping of natural frontiers were not without problems. Bartolo de Sassoferrata, in his treatise *De Fluminibus seu Tiberiadis*, had considered the problems of meanders, changes of river course and new islands in rivers. In his thesis, which was very influential in French judicial geography, he included cartographic solutions to the changes of boundaries caused by alluvial processes.[8] River lines were increasingly used and accepted as frontiers. After the Act of Union of 1707, the shifting channel of the Solway was accepted as part of the Anglo–Scottish frontier.[9]

Yet difficulties arose. In 1719, works on the Elbe led to a serious frontier dispute between Hanover and Prussia. Disputes between the Duchies of Parma and Milan over their frontier on the Po were recurrent, being raised in 1723, 1733 and 1789–90. In 1762, an island in the Ticino led to a border dispute between Milan and Savoy-Piedmont, from 1720 part of the Kingdom of Sardinia. There were also problems with mapping frontiers in mountainous areas.

The eighteenth century

Nevertheless, moves towards more defined frontiers can be discerned in the eighteenth century. They took two major forms, the move towards undivided sovereignty, a crucial precondition for clear territorial mapping, and that towards neat linear boundaries. The former process was handicapped across a broad swathe of Europe by the constitution of the Holy Roman Empire and by such historical legacies as shared authority, for example between the United Provinces and the Prince-Bishopric of Liège in Maastricht, or the alternating control of the Prince-Bishopric of Osnabrück between an elected Catholic prelate and a member of the Protestant house of Brunswick. In the Empire, single maps were not a good way of showing princely territorial rights. As in the modern world, mapping was not well suited to the problems of depicting multiple sovereignty. It was usually beyond the ingenuity of even the most skilful cartographer to indicate on one map areas of mixed jurisdictions, owing allegiance to different rulers for aspects of their existence, for example Schleswig-Holstein.

Considerable success in making sovereign powers more consistent was, nevertheless, achieved. In Alsace, for example, the ambiguous relationship established under the Peace of Westphalia of 1648 between the French crown and the ten Alsatian towns known as the Decapolis was subsequently defined in accordance with Louis XIV's power. The Principality of Orange, an enclave within France, was also acquired by Louis. The move towards frontier lines rather than zones was especially marked in the case of eighteenth-century France, not least because it contrasted so obviously with the complex overlapping of jurisdictional authorities that had earlier arisen because of the impact of French power in the western borderlands of the Empire. This had served French interests as a means of providing opportunities for territorial expansion, most obviously with Louis XIV and the *réunions* in the 1680s, but, in the eighteenth century, *ancien régime* France ceased to be so interested in European territorial expansion, and this new attitude was linked to a desire for the stabilization and, thus, in part, rationalization of frontiers. In 1750, Louis XV ordered Marshal Belle-Isle to settle amicably all the disputes over the frontier between newly acquired Lorraine and the Empire. Some twenty-five years later, the French ministry of foreign affairs gained jurisdiction over boundary matters from the ministry of war, and established a topographical bureau for the demarcation of limits. The context of this stabilization, nevertheless, was one of French dominance over the neighbours who, especially in the first half of the century, complained about the way in which the French used their strength to secure favourable frontier rectifications.

The 'Diplomatic Revolution' of 1756, the Franco-Austrian alliance that lasted, albeit with varying intensity, until 1792, ushered in a period in which French sensitivity over her frontiers diminished and it became possible to negotiate satisfactory solutions to a number of problems. The legacy of border disputes on and near France's eastern frontier was extensive. Peace treaties generally had to be negotiated too speedily for agreements over frontier issues, other than by leaving decisions to commissioners, who commonly found it impossible to settle matters.

The Franco-Austrian alliance can be seen as central to a general pacification of western Europe that was the first pacification of a large area that was in part encoded in maps, thus creating apparently precise borders for posterity. These could be used as a basis for coexistence in future periods of peace, although they also created targets for overthrow and rectification in eras of confrontation and conflict.

In 1749, French frontier differences with Geneva were settled by a treaty, with subsequent delimitation agreements in 1752 and 1763, which was followed in 1754 by the Treaty of Turin between Geneva and Savoy-Piedmont, which had taken since 1738 to negotiate.[10] The French also reached agreements with the Austrian Netherlands (1738, 1769, 1779), the Prince of Salm (1751), the Duke of Württemberg (1752, 1786), the Prince of Nassau-Saarbrücken (1760), Prussia over their common frontier near Neuchâtel (1765), the Duke of Zweibrücken (1766, 1783, 1786), the Bishop of Liège (1767, 1772, 1773, 1778), the Canton of Berne (1774), the Prince of Nassau-Weilburg (1776), the Elector of Trier (1778) and the Bishop of Basle (1779, 1785).[11]

Although older usages persisted, the period also saw the beginnings of a different attitude to frontiers, one that definitely predated the French Revolution. If the right of the strongest still played a major role in the fixing of frontiers, an entirely different principle appeared, that of strict equality between the parties, whatever was the degree of their relative power, both in the course of the negotiations and in the final agreement. The application of this principle was considerably helped by the existence of natural obstacles (rivers and mountains). Using these to draw frontiers offered the possibility of establishing equality in negotiation between the parties in order to create what were termed 'natural' frontiers.

This was certainly the case with the frontiers of Savoy-Piedmont (from 1720 ruled by the kings of Sardinia), and can be seen in the negotiations leading to the convention of 1718 with France, the Treaty of Turin between the two powers in 1760, and the delimitation in the Treaty of Worms of 1743 of a new frontier along the middle of the principal channel of the river Tessin between Austrian-ruled Lombardy and Piedmont. In order to cope with the problems of different channels, the Austro-Sardinian Convention of 1755 and Articles 3 and 9 of the Treaty of Turin both stipulated that the principal channel should be followed and that it should be divided in the middle. Charles Emmanuel III of Sardinia (1730–73), the ruler of Savoy-Piedmont, was not prepared to rely simply on the text of treaties, which were never completely precise. He wanted them represented materially, by a drawing – in other words, a map – and, on the ground, by a continuous demarcation of boundary marks and posts. In 1738, he created the Ufficio degli Ingenieri Topografi. The new Franco-Sardinian frontier was not only 'natural', but also linear. The Convention of Turin of 1760 fixed the Sardinian frontier with Monaco, and that of 1766 with Parma, while the frontier with Genoa was clarified in the Seborga, and between 1770 and 1773 Antoine Durieu and the Genoese Gustavo Geralomo produced joint maps of the hitherto-contested frontier. The frontier with Switzerland in the area of the Great Saint-Bernard was not, however, settled until 1940. The cartographic work of Charles Emmanuel was very extensive: the mapping of frontiers was part of a cartographic tooling of government that included the mapping of his territories. He began with the mapping of the Duchy of Savoy in 1737 and finished with that of his entire territories in 1772.

The same processes of frontier demarcation and mapping were at work elsewhere. Better maps were generally not available for use in international relations until the mid-eighteenth century. The demand–supply question is a subtle one. Diplomats sponsored mapping enterprises, political purposes and values, shaping the content and uses of maps so that a transition from a juridical to a cartographic depiction of boundaries – the former according to lists, the latter according to lines – could take place. In 1743, the Austrians demanded both to see old maps and that the Republic of Venice name an experienced mathematician to work in concert with their own in order to settle the frontier. A treaty was finally negotiated in 1752. A year earlier, there was an important settlement between Sweden–Finland and Denmark–Norway. This laid down the boundary between the two states, ending both serious disputes and Swedish–Norwegian 'common districts' in the interior of Finnmark. The frontier between Russia and Sweden–Finland had also become mappable in terms of linear frontiers. After a conflict ended by the Treaty of Teusina (1595) a frontier had been drawn for the first time between Finland and Karelia – from the isthmus to the White Sea – Russian control of the Kola peninsula and Swedish control over most of Lapland being acknowledged. Nevertheless, Norwegian–Russian 'common districts' – areas of mixed taxation – remained, until they were partitioned in 1826.[12]

Eastern Europe

Maps were used in the protracted and difficult negotiations over the Austro-Turkish frontier that followed the war between the two powers in 1788–91. They eventually concluded a convention that ceded Orsova and a stretch of land on the upper Unna to Austria. The convention, signed on 4 August 1791, referred to lines on a map. The following day Sir Robert Murray Keith, the British mediator, reported:

> The Imperialists had only three copies of the map of the frontiers of the two Empires (which is so often mentioned in the recent Convention); these they have given to the Turks, and to the Prussian Minister. But they have engaged to deliver to each of the mediating ministers, on our return to Vienna, a correct map of that kind with all the limits carefully marked out, according to this last adjustment. I shall think it my duty to send that map to your Lordship, as soon as it shall be put into my hands.[13]

Maps had become important: not as totems, but as the means of diplomacy.

All too frequently, a Eurocentric approach to frontiers and cartography is in fact a western European approach. There were important differences, however, between western and eastern Europe, differences that were to be replicated in (largely western) European imperialist engagement with the non-European world. Within western Europe, a general trend towards more defined frontiers was responsible for some of the warfare in the fourteenth and fifteenth centuries, since lands whose status had been ill defined for centuries were claimed and contended for by rival states. By contrast, the Christian states in the Balkans – Bulgaria, Serbia, even

Hungary – were 'backward' in this respect, and their boundaries remained vague, shifting according to the interests of the local authorities.

Although it might seem that Turkish expansion would have altered the situation, since the Ottoman Empire was a state at least as sophisticated as those in western Europe, in practice the problem remained, because the Turks used 'gradualist' methods of conquest. Christian territories on the borders of lands fully assimilated into the Empire (in the sense that the provincial system of government was implemented) might be compulsorily allied to the Sultan or endure tributary status to the Turks. The territories coming into the latter category were in practice under the Sultan's control: he could move his armies there at will, and demand resources and manpower on the same level as within the Empire proper.[14]

This state of affairs is and was very difficult to depict cartographically. For example, the tributary principality of Wallachia constituted a substantial part of the Ottoman presence in Europe. It would, however, have been, and still be, necessary to give a large amount of detail about relations between the Prince and the Sultan to gain a realistic impression of how powerful the Turks were at any given point beyond the Danube.

Yet, as with European imperialism, there was a powerful factor driving the cartographicization of eastern European international ambitions and power. In western Europe, in the early modern period, the essential unit of diplomatic exchange and strife was jurisdictional-territorial, not geographical-territorial. This was reflected in the dominance of succession disputes in the international relations of the period. For example, most of the major wars in western Europe in the pre-Revolutionary eighteenth century were succession conflicts – the Wars of the Spanish (1701–14), Polish (1733–5), Austrian (1740–8) and Bavarian (1778–9) Successions. The Seven Years War (1756–63) can be seen as an attempt to reverse the principal territorial consequence of the War of the Austrian Succession, and thus as an extension of it.

In eastern Europe, by contrast, geographical-territorial issues played a larger role, and thus encouraged mapping. The major states lacked good historic claims to the areas in dispute, the texture of sovereign polities was less dense than in western Europe – not least because hitherto autonomous regions, such as the Ukraine, were brought under control,[15] and dynastic succession was not the major diplomatic idiom nor generally a means by which large areas of territory changed hands and through which relative power could be assessed. This owed much to the impact of the Turkish advance, while the fact that Poland–Lithuania became a clearly elective monarchy in 1572 was also of consequence. The Habsburgs gained the throne of Hungary by dynastic means, but had to fight the Turks in order to bring their claims to fruition. Succession disputes were not the issue at stake between Russia and its rivals. The idiom of the dispute in eastern Europe was geographical-territorial, and this put a premium on spatial considerations.

Growing distinctiveness of international frontiers

If maps of frontiers were the product of a new sense of international territoriality, that reflected the interest in precision and measurement that was increasingly a feature of European society from the eighteenth century on, and also a growing sense of the distinctiveness of international frontiers. Many European 'frontiers' were, as in Spain after the dynastic union of Castile and Aragon (1479) or Britain after the union of Scotland and England (1603, 1707), essentially domestic-political, most commonly judicial and financial, rather than of any international significance. This process was especially marked in western Europe, with its denser and more historical fabric of jurisdictional authorities, and the accompanying vitality of local privileges. For most Europeans these borders were as significant as their international counterparts, and the two types were difficult to distinguish on seventeenth-century maps. This mental world changed appreciably as the impetus that the French Revolution gave to nationalism from 1789 altered European political consciousness, but, already, prior to that, the increasing demands of sovereign states helped to reconfigure power relationships within their boundaries, thus making the areas comprehended by state frontiers on maps more real as units. Maps, or rather the lines on maps, were more pertinent. The crystalization of European frontiers was, therefore, both real and mappable. The course of frontiers was of interest and could be mapped.

However, there were other types of frontiers that posed different problems. First, there was the delineation of frontiers between European and non-European societies; secondly, the delineation of frontiers between European states in areas where there were no long-standing European settlements. In European eyes, non-European lands could appear empty, non-European societies unsophisticated. The former were appropriated to the European cartographic consciousness either by mapping them as empty, bar a European presence, or by treating them as similar to Europe. Thus Guillaume Delisle's *Carte d'Afrique* (Amsterdam, *c.* 1722) misleadingly divided the whole of Africa into kingdoms with clear frontiers.

Maps were more useful if there was a territorial sense of ownership, a notion of fixed frontiers and a use of natural features as boundaries. These features were cited in treaties between the British North American colonies and Native American tribes, such as that of 1765 with the Lower Creeks in Florida.[16]

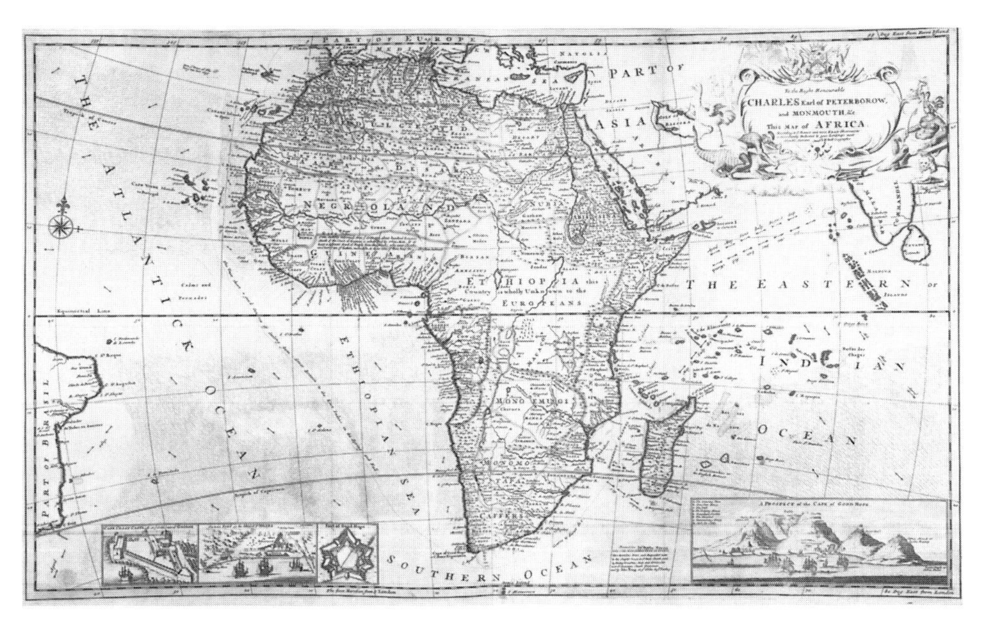

Carte d'Afrique . . ., 1722, by Guillaume Delisle. A European account of Africa in which European notions of territorialization are employed.

Euro-Asian frontiers

The problem of defining frontiers along the Euro-Asian frontier was considerable. The Ottoman Empire had common boundaries with Venice, Austria, Poland and Russia, and from the 1680s Russia 'met' territorially with both China and Persia, leading to border treaties: Nerchinsk with China in 1689 and treaties with Persia in 1723, 1729 and 1732. The Austrian and Turkish commissioners who sought to clarify their new frontier after the Peace of Carlowitz of 1699 faced the ambiguous and contradictory wording of the peace treaty on such matters as the 'straight' line of one portion of the frontier, the 'ancient frontiers', of Transylvania and the future status of islands where the frontier followed rivers. Following the Austro-Turkish war of 1737–9 and the subsequent Treaty of Belgrade, there were lengthy negotiations to settle the new border, and a satisfactory settlement was not negotiated until 1744.[17]

Natural boundaries were an obvious basis for the land frontier between Europe and Asia. There was no jurisdictional definition of territory reflecting long-established political interests. Instead, force operated with scant reference to historical claims. The exact course of this frontier was most important in areas of settlement and along trade routes, and, as these were often riverine, it was rivers that provided the necessary definition, although, by their very nature, rivers are fluid and river courses could vary. For much of the eighteenth century the frontier between Russia and central Asia east of the Caspian – insofar as one can write of one – followed the Ural and Irtysh rivers. The Terek and the Kuban defined much of Russia's frontier in the Caucasus in the late eighteenth century. Further west, the Dnieper, Bug, Dniester and Pruth marked successive stages of Russia's advance across transpontine Europe and towards the Balkans. Similarly, the Oltul, Muresul, Tisza, Danube and Sava played an important role in defining the Austro-Turkish border between 1699 and 1878.

Power and pragmatism divorced from the feudal legacy of jurisdictional issues and non-linear frontiers were dominant on the Euro-Asian frontier. This was certainly the case with the Russian impact in the Balkans, the Caucasus and further east.[18] Elsewhere on the Euro-Asian frontier, the relationship between the coastal enclaves of European trading companies or states in southern Asia and the locally dominant Asiatic powers posed questions of sovereignty and jurisdictional relationship, but the situation was far from uniform. Aside from differences between the imperial organizations and pretensions of different European societies, it was also the case that some Asiatic polities, such as the Indian Mughal and Persian Safavid empires, provided only loose hegemonies, in which it was possible for European interests to establish semi-independent territorial presences, akin to those of some Asiatic regimes.[19] This had obvious consequences in terms of both mappability and the need for mapping.

The New World

Power and pragmatism without the feudal legacy of jurisdictional issues and non-linear frontiers was also the case in North America, where maps played a role in locating and representing territorial disagreements. Maps were employed in Anglo-French differences in the 1680s over the frontier between Canada and the territories of the British Hudson's Bay Company.[20] Differing maps played a role in the failure to settle Anglo-French disputes in North America in 1755. This was related to the fact that the area in dispute was inland, and thus poorly mapped. Attempts after the Anglo-French war of 1744–8 to settle the frontier had been inhibited by the competing views of the two powers as well as bellicose British public opinion. There was a desire for certainty, but a failure to grasp the problems of the issue. Criticizing ministerial pusillanimity, the *Westminster Journal* of 18 March 1749 used a homely image of frontier demarcation when it argued that 'It behoves us to perambulate exactly the boundaries betwixt us and the French in North America; to determine precisely what is ours, and what is theirs, upon the footing of the Treaty of Utrecht.'

However, the scale of the issue was greater then than was suggested by the reference to perambulation: the geographical knowledge of the British commissioners appointed to negotiate frontier disputes with France was limited and their maps were defective. Joseph Yorke wrote of his colleague William Mildmay,

As to Mr Mildmay, I know him very well, and for ought I see he may do very well for one of the commissaries, though it would sure be more decent to nominate somebody, that is more knowing in the geography of America and the West Indies, than I or my supposed colleague.[21]

Conflict in North America broke out in 1754, as a result of competing claims in the Ohio river valley, and in 1755 both powers simultaneously negotiated and armed for conflict. Mirepoix, the French envoy in London, drew attention to the differences between British and French maps of North America and reported being told by Sir Thomas Robinson, the British Secretary of State,

que par l'irregularité du local, la différence de leurs cartes et des notres, et l'infidelité des unes et des autres, il étoit impossible de fixer des lignes justes qui puissent satisfaire aux objets des deux nations et que c'étoit sur cette considération que sa cour proposoit de convenir sur les degré de latitude, la methode la plus seure pour se décider sur des parties aussy peu connues.[22]

In February 1755, the French envoy in The Hague discussed the dispute with Fagel, who was, in effect, the Dutch foreign minister,

La carte de Danville étoit sur ma table. Il l'examina, et aprez un moment de silence, il me dit, 'si les positions sont exactes sur cette carte, il n'y a point de doute que les prétentions de la France, ne soyent légitimes; mais les Anglais peuvent avoir des cartes sur lesquelles il serait peut être aussy aisé d'adjuger la raison de leur côte. Il m'en arrive une d'Angleterre'.

Anglo-French discussions made reference to maps.[23] It is scarcely surprising that the crisis and subsequent conflict created a public market for more maps.[24] New maps of North America were announced in the issues of the *Daily Advertiser* of 3 August, 5 September and 10 September 1755. A *Universal Geographical Dictionary; or, Grand Gazetteer* (London, 1759) was, the title-page proclaimed, 'Illustrated by a general Map of the World, particular ones of the different Quarters, and of the Seat of War in Germany'.

At the end of the war, the *Universal Magazine* of March 1763 provided a map of the extent of territory Britain now controlled in North America. There was also pressure for a clear-cut territorial settlement. 'Nestor Ironside' urged in the *London Evening Post* of 23 September 1762,

let our negotiators take great care, that the bounds of our dominions in all parts of the world, with which the new treaty, whenever it is made, shall meddle, be plainly and fully pointed at; and sure it would not be amiss, if authentic charts or maps were thereunto annexed, with the boundaries fairly depicted. The late peace of Aix la Chapelle [1748] proved indefinite for want of this precaution.

The 1763 settlement was, indeed, much clearer in North America and the West Indies, although there had been problems over the Mississippi boundary in the negotiations; but then the outcome of war had been decisive. A map was joined to the instructions to the British negotiator, the Duke of Bedford, in order to help him negotiate the Mississippi boundary.[25]

These were not the last problems in mapping North American frontiers. The newly independent USA was involved in frontier disputes with Britain and Spain. Owing partly to deficiencies in mapping, the American–Canadian border was drawn in a contradictory manner, leading to a series of disputes that were only finally settled in 1842. Additional disputes arose over the westward expansion of both Canada and the USA, although they were all settled without conflict. For example, the San Juan Islands dispute was settled by arbitration in 1872. The views and interests of native peoples were ignored in such arguments.[26] A settlement of the disputed frontier with Spanish West Florida in 1795 brought America much of the future states of Mississippi and Alabama.[27]

There were also serious disputes over frontiers between individual colonies and later states. Rivers were very important in the creation of colonial, state and county boundaries in the United States, but they could not be used everywhere. Despite tentative agreements between Pennsylvania and Maryland in 1732 and 1739, neither resulted in a permanent solution. In 1763 David Rittenhouse made the first survey of the Delaware Curve, which would not be defined satisfactorily until 1892, and the remainder of the Pennsylvania–Maryland boundary was settled by Charles Mason and Jeremiah Dixon in 1764–7. Violent disputes between Connecticut and Pennsylvania, beginning in 1769, were only settled by Congress's acceptance of the Pennsylvania claim in 1782. The Pennsylvanian General Assembly ordered the survey of the northern line in 1785.[28]

The nineteenth century

Emphasis on frontiers increased in the nineteenth century. This reflected the greater pace, extent and intensity of imperialism,[29] and the role of nationalism within Europe. An understanding and definition of boundaries came to play a major role in public ideology within Europe,[30] one that was expressed in the maps found on so many school walls. Within Europe, boundaries were to be defined; outside Europe they were to be extended. Power was territorial: that was its legitimacy as far as Europeans were

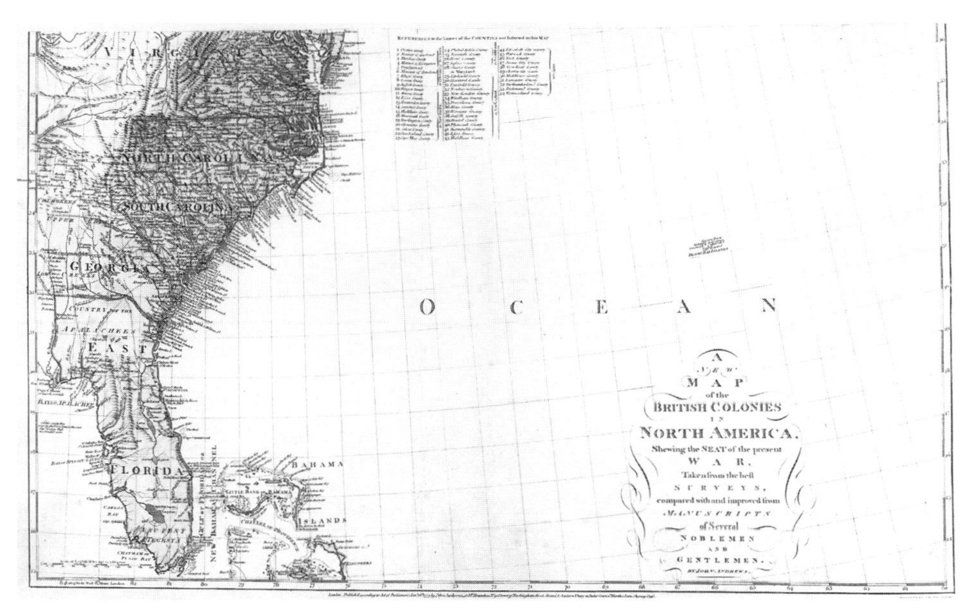

A New Map of the British Colonies in North America, Shewing the Seat of the Present War (1777) by John Andrews. An example of commercial map-making to serve a public avid for news.

concerned. This territoriality required knowledge – locational specificity – and the construction and acquisition of that knowledge was part of a more general process by which the Europeans sought to understand the world in their own terms. This process was not unique to Europeans. For example, provincial boundaries appeared on seventeenth-century Chinese and Japanese maps.

Terminology can convey different meanings here and elsewhere, for example, instead of 'understand' it is possible to use the term 'grasp'. Thus the physical geography of the world was measured – seas charted, heights gauged, depths plumbed, rainfall and temperature graphed. All was integrated, so that the world was increasingly understood in terms of a European matrix of knowledge. Areas were given an aggregate assessment – for example wet, hot, mountainous, forested – that reflected and denoted value and values to Europeans, and these were revealed and indicated through maps.

Territorial politicization was thus but an aspect of, indeed in some respects a stage in, the understanding and allocation of space. It was, however, the most contentious. The world was divided up at a hitherto unmatched rate. In part, this was a matter of the largely uncontentious delimitation of boundaries that had already been essentially agreed. This was true, for example, of the Franco-Spanish treaties of 1866 and 1868 delimiting their Pyrenean frontier.[31]

Frontier delimitation reflected the greater power of the state within Europe and the projection of European power overseas. It was also made possible, but in turn stimulated, by changes in the possibilities of delimitation. Hitherto uncharted waters were sounded. Commander Edward Belcher, in command of HMS *Sulphur*, surveyed the sea around Hong Kong in 1841, the year in which the British took formal possession of Hong Kong island. The two were part of the same process.

More generally, mountains and marshes were explored and mapped around the world, and thus the pressure grew to extend and clarify frontier lines. This was also a matter of scale. First, the quantity of information available for mapping increased. Far more places could be located accurately on distant maps. Secondly, as the world was mapped at progressively larger scales, so it became necessary to seek conformance between lines on the map and the world as refracted through the mapmakers' skill. Outside Europe, this process was generally settled with little reference to natural or ethnic features.

Imperialism and frontiers

The use of lines of latitude and longitude for boundaries was not simply convenient, offering relative precision and speed. It also appeared not simply reasonable, understandable or (even) commendable, but also natural. This was not simply a matter of the delimitation and mapping of colonial territory, especially in Africa. The 49th parallel was used for most of the course of the American–Canadian frontier. Thus knowledge, in this case of latitude and longitude, served as part of the process by which all spatialities were subordinated to a European spatiality; in which morality was provided by its apparent scientific objectivity.

A similar process was involved in another aspect of nineteenth-century knowledge accumulation and depiction in which mapping played a central role: geological surveys. These were very popular in the period. Accompanied by sections, they offered a three-dimensional mapping that appeared to make more sense of the surface of the Earth, to enable it to be grasped in the human mind and for human purposes. Thus John Wesley Powell, the influential Director of the American Geological Survey until 1894, saw geological mapping as fundamental to scientific exploitation: the West was to be classified as mineral, coal, pasturage, timber and irrigatable lands.[32] Knowledge was intended to facilitate and organize more appropriate exploitation, but the sense of 'appropriate' was essentially a matter of maximization of economic benefit. The Earth accounts of other non-Western societies were ignored, marginalized and magicalized – turned into accounts of anthropological interest that reflected the apparent inferiority of these people. This was an aspect of the manner in which the many ethnological maps of the nineteenth century 'fixed' people and themes according to European cartographic preconceptions, as in the racial maps of the period.

Boundaries, however, were not an invention of European imperialists. For example, in the American West, 'the traditional boundaries of the Zunis were known and understood by members of other neighbouring tribes', and it is therefore possible to map the area of Zuni sovereignty in 1846.[33] Nevertheless, concepts of frontier outside the European world were far more multiple than this example might suggest. Many native peoples were nomadic or semi-nomadic: in Asia, Australasia, Africa and the New World. Seasonally occupied areas led to indefinite extent, as, more generally, did social environments based on hunting and/or pastoralism. Whatever the nature of economic activity, identity and political control in many parts of the world was a matter of a people rather than of rule over territory. This could encourage a lack of interest in territorial boundaries or a fluid sense of them.

Furthermore, the politics of frontiers depended on the extent to which neighbouring groups were allies or enemies. They also depended on the nature of the physical geography. For example, it is and was impossible precisely to define the Horn of Africa's frontiers prior to the late nineteenth century, unless there is some prominent and historically important defining geographic feature, such as the Kwarran uplands, the Awash River, the Blue Nile Gorge. Otherwise, the value of drawing a line in the sand or among the trees was limited; yet that was precisely what the European imperial powers chose to do. Their drawing of straight lines on maps without regard to ethnic, linguistic, religious, economic and political alignments and practices, let alone drainage patterns, landforms and biological provinces, was a statement of political control and an exercise of its use to deny existing indigenous practices and assert the legitimacy only of the new, and only of a new that derived directly from European control.

In some parts of the world, the arrival of European power entailed little more than the delimitation of hitherto 'vague' frontiers or the relimitation of boundaries, but, whatever the nature of indigenous spatiality, it was subordinated to imperial cartography. The energy of the latter was expressed not only in the drawing of boundaries and the tensions these reflected or created, but also in the contribution of cartography to the creation of myths that sustained imperialism. Military surveyors played an important role in mapping areas in which European forces campaigned, providing locations for relief and vegetation types, settlements and other features. Surveyors also helped the process of European colonialization, often, as in Algeria and Kenya, an anchor of imperial expansion; and drew attention to mineral deposits and possible trade routes. The maps that were produced for European governments and purchasers ignored or underrated native peoples and states, presenting Africa and other areas, such as Oceania, as open to appropriation. This mapping helped to legitimate imperial expansion, to make the world appear empty, or at least uncivilized, unless under European control. More generally, geography was closely related to several facets of imperialism.[34]

More and more frontiers were demarcated in European terms in the nineteenth century, for example, in the Far East the Russo-Japanese frontier (1885) and in 1861 the Sino-Russian and Korean-Russian frontiers. By contrast, aside from the Gorbitsa River, frontier references in the earlier Sino-Russian Treaty of Nerchinsk of 1689 were quite vague, and their location is still in dispute, with consequences in terms of the Sino-Russian confrontation from the 1960s. Russian, Chinese and Japanese historical atlases project mutually contradictory configurations of historical frontiers in the Far East. Each offers a spatial specificity that assumes geographical knowledge that did not exist prior to the nineteenth century.

Demarcation involved the imperial power or powers seeking to enforce its or their authority in poorly mapped and difficult terrain, as with the Sudan-Uganda Boundary Commission in 1912–13, or the Nigeria-Cameroon Boundary Commission of 1912–13. Most of the boundary lines drawn paid no attention to local identities, interests and views, and this has led to much subsequent criticism: the practices of European imperialism have been blamed for post-colonial ethnic conflict in Africa. However, some efforts were made to avoid dividing tribes. This was certainly true of the Sudan–Uganda Boundary Commission, although its task was eased by British control of both territories.

Demarcation also interacted with local struggles for primacy. The latter had sometimes had a spatial component and these could be incorporated into the territorial strategies and disputes of imperial powers. The Middle East offered good examples of this process. There had long been struggles between powers based in Persia and those further west, especially over the Caucasus, Armenia, the Zagros mountains and Mesopotamia. These lands had been fought over between Parthia and Rome and between the Persian Safavids and the Ottoman Turks. Then, in the early nineteenth century, Britain and Russia came to play a greater role in the affairs of Persia and Turkey. They were transformed from aggressive problems, as Persia had been in the Caucasus in the 1790s, into areas of unpredictable weakness, which offered opportunities for European rivals to gain influence. Disputes over the Perso-Turkish frontier in Kurdistan led in 1843–4 to the formation of a quadripartite Turco-Persian Boundary Commission involving the two powers, Britain and Russia. Extensive negotiations led, despite the reluctance of Persia and Turkey to compromise, to the Second Treaty of Erzeroum (1847) and to an Explanatory Note of 1848 that dealt with ambiguities in the Treaty. The entire land boundary was allocated, although the territorial limit was loosely defined along the east bank of the Shatt al Arab River. It proved difficult, however, to delimit the boundary on the ground, and disputes continued. Indeed, the work of the Delimitation Commission helped to encourage such disputes, which were a serious problem in 1853–60. It was not until 1867–9 that a 'Carte Identique' was completed and a 'Status Quo' Convention signed. Nevertheless, nomadic movements in the 1870s infringed the status quo agreement and disputes continued, for example over Shallah Island in the Shatt al Arab.

In response to continuing problems, the Russians proposed to reconvene a boundary commission, and in the Teheran Protocol of 1911, Persia and the Turks agreed that a new commission should begin work, based on the 1847 treaty. Meetings in 1912 proved fruitless, but in 1913 the British obtained Turkish agreement to a boundary line that led to the Constantinople Protocol of that year, which clarified the delimitation. The Frontier Commission continued its work until World War One.

After the war the situation deteriorated anew because of border disputes and because the successor governments in Persia and Turkey were unwilling to accept previous compromises. Kemal Ataturk's nationalist government in Turkey rejected the validity of the 1913 Constantinople Protocol. The issue became more serious in the 1920s as oil exploration increased in the region, and in the early 1930s Iran (Persia) and Iraq (the successor state in the region to the Ottoman Turkish Empire) accused each other of frequent border violations. The League of Nations failed to settle the dispute in 1934–5, but in 1937 Iran and Iraq signed a boundary treaty that temporarily appeared to do so. However, the Iran–Iraq Frontier Commission was a failure and disputes over the Shatt al Arab waters resumed.[35]

Thus, during the period of imperial power and influence, other powers, such as Persia, Turkey, Iraq and China, had to accept the consequences of European concern with frontiers. This concern, however, interacted with often longstanding local and regional disputes. This can also be seen by British intervention in boundary issues in the Arabian Peninsula. These again involved independent rulers, such as Turkey, Saudi Arabia, the Imamate of Yemen, Kuwait, Muscat and the Gulf states. The British played a major role in establishing and delimiting boundaries. In part, this was a matter of ensuring that hostile powers were kept at a distance: thus Turkish claims and expansionism were resisted in 1879–1914. Secondly, the British sought to ensure that their own territories were clearly marked.

The frontier of the Imamate of Yemen and the Aden Protectorate was delimited in 1903–5, and the boundary settlement of 1905 ratified in 1914. A procedure for the settlement of frontier disputes was agreed in 1942. The boundary between Saudi Arabia and the eastern Aden protectorate, however, led to disputes and extensive negotiations.

In Africa, interest in economic development led to a major increase in mapping in the last decades of imperial control. Similarly, economic interest was important in the European mapping of the Middle East. The search for oil there was a direct response to Western interests, and entailed a new form of imperialism that led to fresh pressure for territorialization. The precision of oil concession areas was imposed on a desert society where the movement of bedouin and their flocks had led to a less territorially fixed understanding of boundaries. This territorialization was certainly the case in negotiations between Saudi Arabia and Britain over the Saudi-Trucial Coast frontier in 1934–44, and again over the frontier between the former and Abu Dhabi in 1947–57. The British urged the Sultanate of Muscat and Oman, a client state, to claim a western boundary with Saudi Arabia in the 1940s, and this, in turn, exacerbated Saudi–Muscat relations. In 1954, the dispute between Abu Dhabi and Saudi Arabia over the Buraimi oasis became more serious because of the issue of oil concessions.[36] In 1954–7, the British arbitrated the borders of the Northern Trucial States, and in 1957–60 most of those between Oman and the Trucial States.

Similarly, in the Persian Gulf, oil made issues of control over islands and seabed more serious from the mid-1930s.[37] The definition of Bahrain's offshore oil concessions became an issue. Furthermore, in 1938, oil companies enquired about the status of islands in the Gulf. Drilling began in submarine parts of the Gulf. In 1949, Saudi Arabia issued royal proclamations extending territorial waters to six miles and claiming ownership over the resources of the seabed and subsoil of the continental shelf beyond these waters. Britain responded on behalf of Kuwait, Bahrain, Qatar and the Trucial Coast states; claimed areas were rapidly allocated. In 1949–50, offshore oil concessions were granted to the Superior Oil Company by Qatar and Dubai and to the American Independent Oil Company by Kuwait. When Shell acquired the Qatar offshore concession in 1952 the issue of 'safe' operating limits arose, and this caused difficulties with Bahrain in 1955, when Shell wished to drill up to the western limit of the Qatar offshore concession area.[38]

The role of oil reconfigurated imperialism and power relationships, and ensured that longstanding issues of European versus local concepts of territory and criteria for sovereignty became of crucial economic importance. Thus a greater degree of attention was devoted to them in particular contexts than would otherwise have been the case. European assumptions prevailed; for example, Britain and the Ottoman Empire had reached agreement on their respective spheres of influence in Arabia, drawing boundaries known as the Blue (1913) and Violet (1914) Lines. After the Ottoman Empire collapsed, Britain claimed that these lines devolved upon the successor states of the area, and continued to insist on their validity, ensuring that these lines and their amendments became the basis for disputes between Saudi Arabia and its neighbours.[39] The international bodies to which reference was made, such as the International Court at The Hague, the League of Nations and the United Nations, similarly enforced European territorial norms.

Oil frontiers did not play a role in the territorialization of Palestine, but, again, European norms came to prevail. British determination to protect the Suez Canal led to a demarcation of the Egyptian–Turkish frontier on British terms in 1906. The notion of a League of Nations mandate was the basis of European territorial control of most of the Middle East after World War One, and British policy was influenced by Zionism, a transplanted and transplanting European nationalist movement. Borders between communities and frontiers between states were to be a longstanding cause of tension in the region and this affected its cartographic history. The contentious nature of the ownership of land and of water rights ensured that records were sensitive and attempts were made to destroy them or otherwise to control the record. The mapping of historical episodes was intended to affect the legitimacy of current territorial boundaries. Thus Martin Gilbert, a keen supporter of Israel, in his *Atlas of the Arab–Israeli Conflict* (6th edn, London, 1993), mapped the McMahon–Hussein correspondence of 1915, under which the British High Commissioner in Cairo had proposed that districts to the west of Aleppo, Hama, Homs and Damascus should be excluded from a proposed independent Arab state on the grounds that they were not purely Arab. Zionists seized on this to argue that Palestine was not therefore included in the state, and hence could be settled by Jews, but their interpretation was hotly disputed by Arab leaders, who argued that the wording referred to Lebanon alone, where there was a substantial Christian population. Gilbert draws his map to exclude all of Palestine from the projected Arab state, although he notes that no mention was made of Jerusalem, the Jews or southern Palestine in the correspondence.[40]

The Europeanization of world political space and spaces in the nineteenth century was maintained in conceptual terms in the twentieth, but was refracted through different political circumstances, particularly the collapse of empires, both in Europe and elsewhere, and the accompanying rise of national-ethnic logics of territorial space. Thus, in Europe, the collapse of the German, Austro-Hungarian and Russian empires at the end of World War One, led, after the Paris Peace Conference of 1919, in which geographers played an important advisory role, to a period of frontier drawing in which local consent, assessed by plebiscites, played a role. For example, the Danish–German frontier was settled by a plebiscite in 1920, creating a frontier that has since remained unchanged.

Such consent-frontiers were not granted outside Europe: the victors and the League of Nations introduced and maintained very different logics of territorial legitimacy outside and within Europe. Plebiscites were not used. Thus, for example, in 1939, France, which held the League of Nations mandate for Syria, accepted Turkish claims to the Alexandretta region, creating the basis for a postwar dispute between Syria and Turkey. In the 1990s, Syria still includes Alexandretta as part of the country on its maps. British strategic interests delayed agreement on the Transjordan-Iraq frontier until 1932, and a border survey was not conducted until 1940.

Within Europe the defeated powers also suffered territorial losses that they found difficult to accept. Maps served to make vivid a sense of territorial loss. Budapest had a memorial garden laid out as a map of the old kingdom. It both recorded the losses suffered by the Treaty of Trianon of 1920 and made them appear unnatural.

Post-1945

After World War Two, the European colonial empires collapsed, although the process took three decades; and in the early 1990s the Soviet empire also collapsed. Since 1945, over 120 new states have been created. Decolonization, whether peaceful or not, immediately led to new international frontiers and to disputes, for example the post-Soviet conflict over offshore boundaries in the Baltic and Caspian Seas by the new littoral states. It also ensured that earlier disputes that would have been limited by the strength and influence of imperial powers, whether as participants or as arbiters, became more urgent. This was true for example of the Anglo-Argentine conflict over the Falkland Islands, of the border dispute between Latvia and Lithuania over oil-exploration rights in the Baltic Sea, and of the disputes in the South China Sea. Such disputes led to a cartography of claim that was especially marked in states seeking to reverse existing territorial arrangements, for example Ecuador, which was unhappy with its trans-Andean frontier with Peru after the 1942 clash between the two powers. This affected maps such as those in the *Atlas Historico-Geográfico del Ecuador* (Quito, 1990), a work produced by a state body, the Instituto Geográfico Militar. Imperial states, especially the Soviet Union, had not always demarcated and mapped internal borders adequately, and this created serious problems when, with decolonization, they became international frontiers.

Aside from the cartography of claim on the part of states dissatisfied with their frontiers, more generally, autocratic states encouraged an enhancing cartography. Modern Chinese maps exaggerate the extent of past Chinese empires,[41] and thus lend weight not only to specific territorial pretensions, but also to a more general sense of potency. This sense of potency can be communicated in a number of ways in maps and atlases. The emphasis on unity present in the use of national units for maps can be underlined in the text, as in the *Atlas de Madasgascar* (Tananarive, 1969), for which the Foreword was written by the President. Claims to an identity between people and territory can be asserted through maps and extended back through time. Thus, Dacia, the Roman Empire north of the Danube, was divided in the maps of the modern *Tabula Imperii Romani* project between sheets produced in Hungary (L34, Budapest, 1968) and Romania (L35, Bucharest, 1969), and their different impression of the 'Romanness' of the region reflected controversy between the two states over Romanian claims for ethnic continuity since the Roman period.

Welcomed as progressive, self-determination, like decolonization and nineteenth-century nationalism, was another cause of international instability and cartographic uncertainty. The principle of self-determination failed to address the issue of who was allowed to seek it. In 1960, the United Nations stated that all 'peoples' had the right to self-determination,[42] but it was, and is not, clear how 'peoples' were to be defined.[43] 'Peoples' can be constructed as much as nations. For example, the Ovimbundu of Angola consist of a dozen warring tribes and not a single 'polity', as the public-relations advisers to the UNITA (National Union for the Total Independence of Angola) leader Jonas Savimbi claim. Indeed, the failure to focus on this issue is a challenge to much writing about nationality and nationhood, and a major problem for the political mapping of geopolitics and 'real' boundaries.[44]

The assertion of identity encourages mapping. Some is uncontentious; the wallmap *Robert the Bruce: Maps of the War of Scottish Independence and the Battle of Banockburn* (Edinburgh, 1974) applauded Bruce and the gaining of independence in the fourteenth century at a time of rising nationalism in the twentieth. In other contexts, especially those directly addressing contemporary issues, such maps are more contentious. The Turks do not encourage maps of greater Kurdistan or Armenia.

Boundary disputes after 1945 did not only reflect decolonization and the struggles of 'peoples' seeking to create new polities or to alter existing frontiers; there were also disputes between states that were long established, or relatively so. For example, the 1867 Convention by which the

USA had purchased Alaska from Russia defined a maritime boundary that was subsequently interpreted differently by the USA and the Soviet Union, leading to different maps. It was not until 1990, and then only after nine years of negotiations, that the two powers signed a maritime boundary agreement fixing the longest maritime boundary in the world and one that could be shown in an agreed fashion on the maps of both powers.

Territoriality and frontiers

Apart from in North Asia, where Siberia remains Russian, direct European political control elsewhere in the world was less in 1997 than had been the case for a half millennium. Yet European notions of territoriality have been assumed and internalized by non-Europeans,[45] and colonial divisions of peoples have been maintained. For example, the Anglo-French Treaty of 1898, which divided Hausaland and which had led to a joint commission in 1906–8, has been maintained as the boundary between Niger and Nigeria, and the impact of colonialism is still felt in local government, education and economic practice. The Hausa are now defined by the impact of a frontier that is no longer alien.[46] More generally, the different values often advanced in neighbouring states ensure that frontiers can also mark important psychological boundaries.

The frontiers of imperial and colonial power and control are now challenged by native peoples using the law courts in order to assert traditional claims, but they still have to do so in 'Western' territorial terms. Thus, for example, the Musqueam have filed a land claim for much of Vancouver, although such claims are not shown in 'Western' atlases. *Vancouver: A Visual History* refers to the Musqueam claims in the text, but does not show them in the relevant map.

International frontiers therefore are not important solely to the world of diplomacy; they also define and influence the local contexts of life. Maps mark and confirm this territorialization and are crucial to the process by which it is made to seem natural, a form of knowledge made concrete. The shapes of countries or regions, such as France or California, on maps become familiar through repetition in particular perspectives, and, being familiar, seem natural. However, with the exception of the world, a special case, it is more difficult to make the cartographic image of a supra-national entity seem natural. This is true of obvious man-made constructs, such as the European Union.

There are also problems in fixing the cartographic image of many entities that have a natural dimension, such as the Baltic lands, especially if these entities have been fractured by political hostility:

Many a Finnish or Swedish coastal dweller remembers the recent time when contacts with Estonia, Latvia and Lithuania were so slight that in their mental picture the Baltic seemed to continue eternally, like an ocean, from Helsinki or Gotland, while those on the opposite shore remembered and pined for the world beyond the Baltic in another way, although education and official ideology urged them to turn their eyes inland. The change in the political, economic and cultural situation has also changed the landscape; the 'opposite shore' has become much more familiar thanks to tourism and business, as the mental picture of the landscape has also changed.[47]

Thus the Cold War frontier inhibited the mental mapping of the Baltic and also ensured that maps of it as a geographical space created an erroneous impression. The frontier as line on the map can be an insufficient statement of borders and boundaries, particularly of their weight, impermeability and mental impact.

The Baltic also provides an example of the manner in which the inclusion of particular territories and states in maps purporting to cover an entity, such as Europe or, in this case, Scandinavia, reflects political aspirations and perceptions. The Russian acquisition of Finland from Sweden in 1809 was followed by the area being omitted from maps of Scandinavia and, instead, generally included on the northwest corner of maps of European Russia, as part of the Baltic provinces of the Russian Empire. The gaining of independence by Finland, Estonia, Latvia and Lithuania at the end of World War One led to their being included in maps of Scandinavia, but after 1940, when the Soviet Union annexed the last three, only Finland was thus included. The Finns were sensitive to their cartographic identity, producing an atlas of Finland in 1899 and their first atlas as an independent state in 1925. Concern about Finland's status as a European country was further reflected in the exhibition 'Finland: 500 Years on the Map of Europe' that the Ministry for Foreign Affairs 'commissioned' from the Museum of Central Finland in 1993. This was designed, as the catalogue noted, to display 'through the information provided by the evolution of maps, how Finland developed from a sparsely inhabited and remote Swedish province into a developed European state'.[48] Russian and Soviet cartographic views were ignored.

Conclusions

Many state boundaries are relatively recent – under 150 years old. Cartography was an important tool in staking and sustaining territorial claims during the process of their definition and defence. Boundaries have been generally more stable in the last fifty years than they were in the preceding sixty, a process that has owed much to the extent to which decolonization proved less disruptive of frontiers than colonization had done. As a consequence, the role of maps in asserting territorial claims has been less familiar than it was, for example, in Europe in the first twenty-five years of the twentieth century. This may change as the legitimacy of states is challenged from above and below in a process of regionalization. Changes in Germany and the former Yugoslavia and the Soviet Union indicate geopolitical fluidity, while in the case of Yugoslavia maps have been used to assert ethnic identities and territorial claims, and to plan a territorialization based on 'ethnic cleansing'.[49]

Frontiers assert and separate identities. They reflect and create borders, and borders produce their own geography.[50] The existence of frontiers encourages mapping, and they are in large part known through maps. The map as expression of state power and the map as the creator and sustainer of images of national identity and shape, coincide and interact at frontiers. So too does the map as tool of war.

6 War as an Aspect of Political Cartography

If frontier disputes require cartography and increase governmental and public interest in it, the same is even more true of war: maps can serve to define mutual boundaries, but they are also indicators of the processes by which such definitions prove elusive or cannot be sustained. War, the pursuit of politics through the overt and violent application of organized force, is an exercise in the location and exertion of force, and requires both a close understanding of the spatial dimension of force and a control of territory itself. Maps are operational, required for strategy and tactics, communications and logistics. They define and clarify the territorial aspect of conflict, although other aspects of war cannot be mapped. Morale and, more generally, relevant cultural suppositions, such as the willingness to bear heavy casualties, cannot be indicated, and there are also problems with the depiction of command and control capabilities.

The military applicability of maps also depends on the particular characteristics of the weapons systems being employed, and maps of war can only be understood if the operational capabilities of these systems, and of opposing devices, are grasped. Thus, for example, the mapping of war in a given area at the close of the twentieth century would be different if only surface conflict is to be shown or if the aerial dimension is to be included. Again, the nature of the aerial dimension varies if the maps are to include air-to-surface operations and surface-to-air responses, such as anti-aircraft fire, as well as air-to-air conflict.

Much of the history of European cartography centres on its military rationale and application,[1] and much of the cartography was prepared under military aegis, or for military purposes. For example, most early maps of the Plymouth region were drawn for reasons of defence. Military concerns also led to mapping by non-Europeans. A tradition of cartographic reconnaissance developed in the Ottoman army in the the second half of the fifteenth century, and numerous forts were mapped in South Asia.[2]

Within Europe, the skill base required for cartography was not restricted to the military, but it was the military that had the ability, resources and need to survey and map large areas at various scales. Thus, in 1747–55 the British surveyed the mainland of Scotland at the scale of 1:36,000 in order to produce a map that would, it was hoped, enable the army to respond better to any repetition of the Jacobite rising of 1745. This was the cartographic equivalent to the road and fortress-building policies of the same years. Thus Scotland, more specifically the Highlands, was to be controlled in a variety of related ways. Fortresses anchored the governmental position at crucial nodes, roads radiating from them offered approaches into the Highlands, and maps provided guidance in the planning and use of force. Maps also offered the prospect of a strategic, Scotland-wide, response to any future Jacobite uprising.

At the scale of global history, European maritime hegemony from the sixteenth century rested, in part, on cartographic developments in Europe that permitted the depiction of the world's surface on a flat base in a manner that encouraged the planned deployment and movement of forces. This was due to 'the rediscovery of the geometry of the Greeks, the discovery of the laws of perspective, the development of abstract thinking and science, and the development of a more abstract language, the introduction of paper as drawing base, the introduction of printing'.[3] Thus cartography was a crucial aspect of the ability to synthesize, disseminate, utilize and reproduce information that was crucial to European hegemony. The movement of ships could be planned and predicted, facilitating not only trade but also amphibious operations. Maps served to record and replicate information about areas in which the Europeans had an interest and to organize, indeed centre, this world on themes of European concern and power.

War, furthermore, encouraged public interest in mapping. Reviewing Lewis Evans' *Analysis of a General Map of the Middle British Colonies in America*, in 1756, Dr Johnson wrote that 'the last war between the Russians and the Turks [1736–9] made geographers acquainted with the situation and extent of many countries little known before';[4] a reference to the lands on the northern shore of the Black Sea. This continued to be the case. Conflict encouraged both supply and demand – military mapping and the commercial production of maps. In 1776 the London map-seller Carrington Bowles published a *Map of the Seat of War in New England . . . Together with an Accurate Plan of the Town, Harbour and Environs of Boston.* The map was sufficiently detailed to enable British readers to follow the course of the conflict. The same year, a more detailed map of the Boston area showing the location of British and American positions was also published in London. Furthermore, the battle of Bunker Hill in 1775 was rapidly followed in London by the publication of maps of the battle, the first appearing four days after the report of the engagement reached London. The *Journal Politique de Bruxelles* of 2 February 1788 advertised

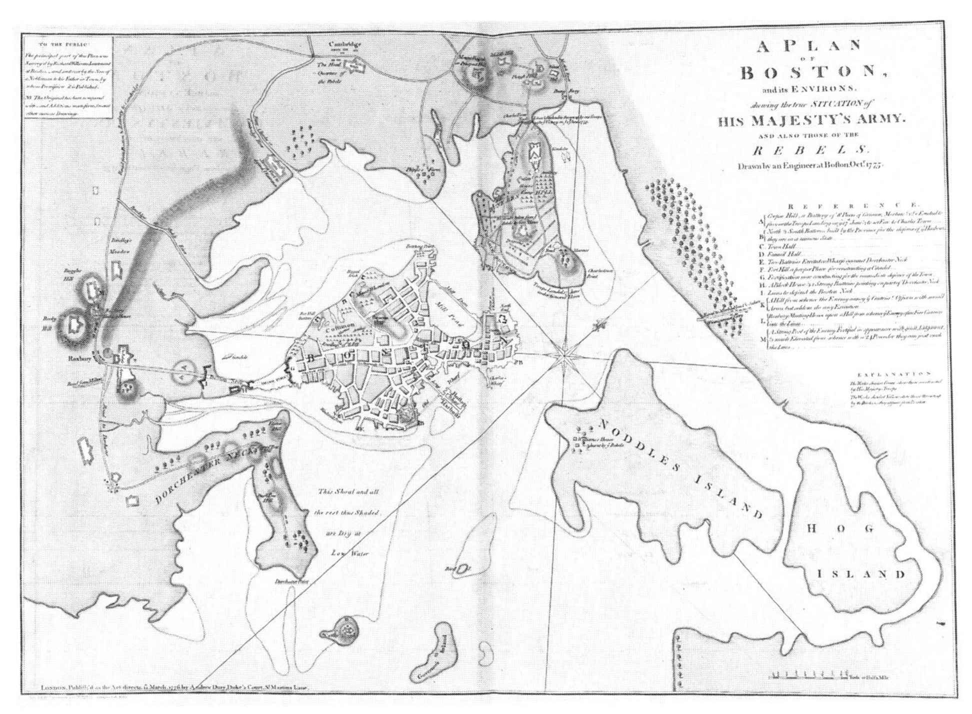

Anonymous *Plan of Boston and its Environs, Shewing the true Situation of His Majesty's Army. And also those of the Rebels* (1776). Foreign campaigns greatly increased the number of maps that were publicly available. The British public was offered more maps of their American colonies during periods of war than in those of peace.

a map of the northern and northwestern littoral of the Black Sea that would help those interested in the recently commenced Russo-Turkish war to follow its course. The British government certainly had need of reliable maps of the Black Sea, for a lack of information emerged as a problem during the Anglo-Russian Ochakov Crisis of 1791.

The lengthy warfare of the French Revolutionary and Napoleonic period (1792–1815) further encouraged the military production and printing of maps. The British Ordnance Department mapped the British Isles. The Duke of Wellington used a mobile lithographic printing press in the Peninsula War, an example of the military employing new technology. The warfare also encouraged public interest in maps of wars and battles. In the nineteenth century, military production was increasingly institutionalized and became more frequent in both peacetime[5] and war. This was true of the USA as well as Europe. For example, the US Corps of Topographical Engineers, which existed from 1838 until 1862, played a major role in the exploration and mapping of the American West. They also played an important role in providing maps for the war with Mexico in 1846–8.[6]

Military surveyors used Native American maps and geographical information, but only to a limited extent; they were concerned to locate, understand and utilize the lands they surveyed in the context of an expanding continental state: localities acquired meaning with reference to this project. Thus, for example, a major effort was made to decide which routes should be used in creating railways from the Mississippi to the Pacific, as with Lieutenant Amiel Whipple's survey of a possible southern route in 1853–4. The resulting Pacific Railroad Surveys were published as Senate Executive Documents.[7]

The non-military, commercial cartographic world was able to supply maps of spheres of conflict, for example North Italy, but also had to meet a growing public interest in the details of campaigns and in battles. A desire for information and for precision was expressed by the growing political public of nineteenth-century Western states. The literacy of this public was increasing and so too was its interest in politics, both nearby and distant: nationalism and imperialism combined to this end. Thus, whereas a French newspaper reader in the 1790s did not expect to see any maps in his newspapers, despite France's engagement in a bitter war, his descendants a century later wished and indeed expected to see maps recording France's imperial advance, whether in west Africa, Madagascar or Indo-China. These interests provided opportunities for publishers and led to some innovative cartography. For example, in 1855 Read and Co of London published *A Panoramic View of the Seat of War in the North of Europe*, a map that adopted an aerial perspective on Anglo-Russian hostilities in the Baltic. The same year, the British Secretary of State for War founded a Topographical Department.

The American Civil War (1861–5) was a conflict in which government and public demand for information affected mapping, not least by creating an immediacy of public mapping. Field commanders used maps extensively, although, certainly at the outset, they were affected by a shortage of adequate maps. Commercial cartography could not serve military purposes, and thus the armies turned to creating their own map supplies, part of a more general shift between commercial and military sources that affected military cartography in the nineteenth and, even more so, twentieth centuries. By 1864, the United States Coast Survey and the Army's Corps of Engineers were providing about 43,000 printed maps annually for the Union's army. In that year the Coast Survey produced a uniform, 10 mile to the inch base map of most of the Confederacy east of the Mississippi. Technology served the cause of war, lithographic presses producing multiple copies rapidly. The production of standard copies was crucial, given the scale of operations, especially the need to coordinate forces operating over considerable distances. This was true not only of campaigns, but also of battles. The scale of the latter was such that it was no longer sufficient to rely completely on the field of vision of an individual commander and his ability to send instructions during the course of the engagement. Instead, in a military world in which planning, and staffs specifically for planning, came to play a greater role, maps became far more important.[8]

In the American Civil War, surveyors and cartographers were recruited for the military, creating problems for private producers, who were also affected by the impact of military demands on the availability of paper, cotton, fabric, boards and glue. Government intervention could be more direct: some maps were withheld from sale because they were seen as of potential value to opponents.[9]

Newspapers were expected to provide maps, and were able to do so because of the presence of military correspondents and recent advances in production technology in printing and engraving: steam-powered rotary printing presses and wood-engraving. The development of illustrated journalism paved the way for the frequent use of maps. Military correspondents sent eyewitness sketches that were rapidly redrawn, engraved and printed. In addition, publishers issued a large number of sheet maps. The scale of production was vast: between 1 April 1861 and 30 April 1865 the daily press in the North printed 2,045 maps relating to the war. The public need for maps was also served by the publication of large numbers of free-standing maps, especially in the North. These included theatre of war and battle maps, such as that of the Potomac area, which enabled those 'at home' to follow the movements of relatives and others. Conversely, in the South, there was a severe shortage of press operators, printers, wood-engravers and printing materials, all of which combined to ensure that the appearance of maps in southern newspapers was rare, although some, such as the *Charleston Mercury* and the *Augusta Constitutionalist*, did print a few.

Newspaper publication of maps of the conflict led to governmental action. On 4 December 1861 the front page of the *New York Times* carried, 'The National Lines before Washington. A Map exhibiting the defences of the national capital, and positions of the several divisions of the Grand Union army'. The accompanying text began,

> The interest which attaches to the military operations of the National army on the line of the Potomac, has induced us to present the readers of the Times with the above very complete and accurate map of the impregnable lines on the Virginia side of the National Capital ... The principal permanent fortifications, which the rebels, if they attempt them, will find to be an impassable barrier to their ambitious designs upon the Capital, have been enumerated by title and position in the General Orders of General McClellan, but are, for the first time, located and named upon the present map ... Another novel and useful

trait of our present map is its geographical definition of the territory occupied by each of the eight divisions constituting the grand defensive army.

The army's commander, George McClellan, furiously demanded that the paper be punished for aiding the Confederates. The Secretary of War restricted himself to urging the editor to avoid such action in the future. The following spring, however, the War Department established a voluntary system to prevent journalists with the Army of the Potomac from publishing compromising maps.[10]

Mapping of and for war also increased elsewhere in the world. General staffs, organized on the Prussian model, studied past campaigns, conducted manoeuvres and planned for war, as with Schlieffen's 1905 plan for an attack on Belgium and France. All required detailed maps, especially maps that integrated topography and transport routes. However, such maps could mislead, because they provided no guidance as to the impact of climate and weather. This was to be especially serious in areas where roads were turned into quagmires by rain, as in Russia in the spring and autumn. General staffs proved more adept at calculating rail capacity than road capability.

War also increased public interest in maps. The American National Geographic Society first included a map with their magazine in 1899, when America's conquest of the Philippines from Spain the previous year led to the inclusion of a map of the islands. The conflict had already led to the publication of two atlases: *The St Paul Spanish–American War Atlas* (St Paul, Minnesota, 1898), and Arista C. Shewey's *Shewey's Official Handy Reference Pocket and Cyclopedia, containing Authentic Historical Information and Statistical Tables of Reference relating to the Spanish–American Conflict, with official maps* (Chicago, 1898). As imperial armies operated in distant parts of the world, so the public at home expected maps of, for example in the 1890s, Sudan and South Africa. Atlases of recent wars were published, including the *Atlas histórico y topográfico de la guerra de África en 1859 y 1860* (Madrid, 1861) and the *Atlas histórico da guerra do Paraguay* (Rio de Janeiro, 1871).

Given that maps can be seen as a crucial adjunct of power, their direct use for aggressive purposes appears particularly appropriate. In the twentieth century, however, war has inhibited civil mapping, as well as destroyed map-making facilities. For example, the Geological Survey of Ireland's field programme was severely curtailed during the War of Independence (1919–21) and the Civil War (1922–3), because the surveyors were seen as agents of government.[11]

Yet war also led to improvements in the tools and methods of mapmaking, and to developments in depiction. Accurate large-scale maps

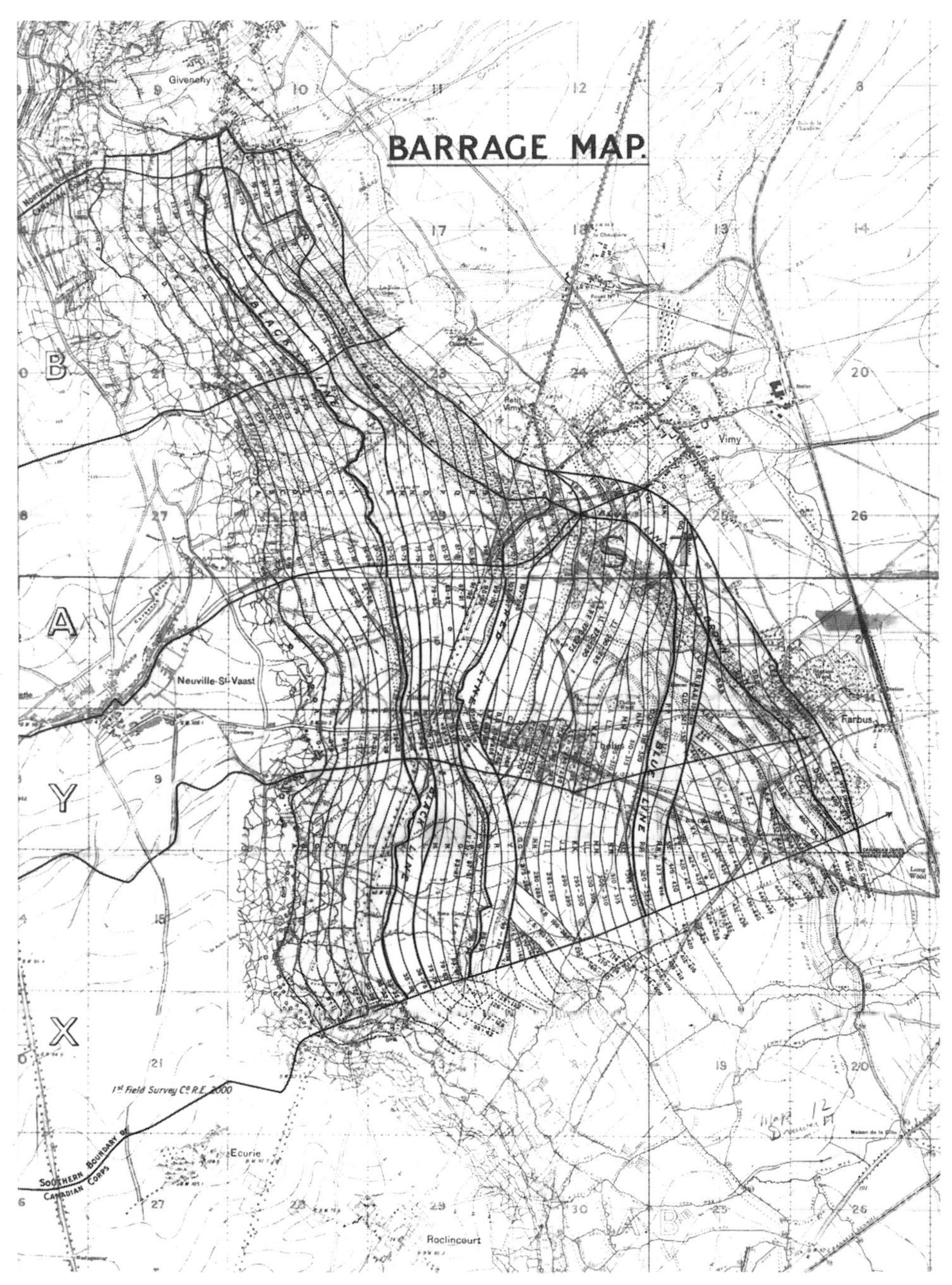

World War One artillery barrage, Vimy Ridge. Trench warfare led to a massive growth in military map-making. Infantry-artillery coordination made accurate coordination crucial.

were crucial for the trench warfare of World War One, not least for artillery locating their targets. The growing emphasis on indirect fire had important cartographic consequences. Trench warfare led to a massive growth in military map-making. In part, this depended on new technology, with an extensive use of aerial photography: cameras mounted on balloons and aeroplanes. There was also a great expansion in the production of maps. When the British Expeditionary Force (BEF) was sent to France in 1914, one officer and one clerk were responsible for mapping, and the maps were unreliable. By 1918, the survey organization of the BEF had risen to about 5,000 men and had been responsible for more than 35 million map sheets. No fewer than 400,000 impressions were produced in just ten days in August 1918. Geographical and cartographic talent was used for the war effort.[12]

The factors of scale were even greater in World War Two. The role of air power, at land and sea, ensured that many maps had to be more complex than hitherto, and there was a premium on cartographic expertise. Thus, for example, Armin Lobeck, Professor of Geology at Columbia, produced maps and diagrams in preparation for Operation Torch, the American invasion of French North Africa in 1942. Quantity was also important. The British Ordnance Survey produced about 300 million maps for the Allied war effort; its offices in Southampton were badly bombed by the Germans in 1940.[13] The American Army Map Service produced over 500 million maps.

Furthermore, war increased the official and journalistic demand for maps. Publishers found that there was a massive market for atlases of World War One: the mobilization of national resources – the nation at arms – and the intensity of the conflict led to a high degree of public interest. This was satisfied by newspaper and book publishers with works such as the *Atlas of the European Conflict* (Chicago, 1914), the *Daily Telegraph Pocket Atlas of the War* (London, 1917), *Géographie de la Guerre* (Paris, 1917), *From the Western Front at a Glance* (London, 1917), *Petit Atlas de la Guerre et de la Paix* (Paris, 1918) and *Bretano's Record Atlas* (New York, 1918).

When World War Two broke out in Europe maps of the Continent sold out in the United States, and the public use of maps increased at the New York Public Library. From 1940, government agents used the collections of the American Geographical Society, while geographers played a role in the British war effort.[14] A shortage of maps in military collections when the United States entered the war in 1941 led to extensive government borrowing of maps, for example from the New York Public Library, and to much microfilming of maps. Walter Ristow, the Chief of the Map Division at the New York Public Library, became head of the Geography and Map Section of the New York Office of Military Intelligence. The government not only sought to acquire maps; it also restricted their distribution, especially some US Geological Survey topographic maps and US Coast and Geodetic Survey nautical charts. Governments also produced maps for public consumption, the American Government Printing Office issuing a Newsmap series.[15]

The war also led to an enormous increase in press cartography. *The New York Herald-Tribune*, *New York Times*, *New York Daily News*, *Christian Science Monitor*, *Chicago Tribune* and *Milwaukee Journal* all had their own cartographers, and their maps were also reproduced in other newspapers. Crucial figures included Emil Herlin, whose maps were reproduced in *The War in Maps, an Atlas of New York Times Maps* (1942), Robert Chapin Jr in *Time*, whose war maps were also separately published by the magazine in much enlarged versions, and Richard Edes Harrison, who had introduced the perspective map to American journalistic cartography when he had produced maps to explain the Ethiopian war in 1935. The preface to Harrison's *Look at the World, The Fortune Atlas for World Strategy* (New York, 1944), explained that it was intended 'to show *why* Americans are fighting in strange places and *why* trade follows its various routes. They emphasise the geographical basis of world strategy'. German firms such as Justus Perthes and Ravenstein produced detailed war maps for the German public.[16] Maps and map projections greatly affected the public image of the war. Roosevelt's radio speech to the nation on 23 February 1942 made reference to a map of the world in order to explain American strategy. He had earlier suggested that potential listeners obtain such a map, leading to massive demand and also to increased newspaper publication of maps.

The map-makers faced a number of problems, some of which related to the general difficulties of depicting war, others to the particular nature of their audience. In the latter case it was necessary to locate and make clear what were to Americans often very distant and unknown regions. It was also important to counter isolationism by linking these regions to American interests: interest and interests had to be focused. Map-making thus served a political end: it was part of a worldwide extension of American geopolitical concern and military intervention. World War Two also globalized American public attention, and maps played a major role in this process. For example, Harrison's orthographic projections and aerial perspectives brought together the USA and distant regions. The dynamic appearance of many war maps, for example those in *Life* and *Time*, with their arrows and general sense of movement, also helped to convey an impression that the war was not a static entity at a distance, but, rather, was in flux and therefore could encompass the spectator both visually, through images of movement, and, in practice, by spreading in his or her direction.

The role of air power, dramatized for Americans by the Japanese attack on Pearl Harbor in 1941, led to a new sense of space, which reflected both vulnerability and the awareness of new geopolitical relationships. The Mercator projection was unhelpful in the depiction of air routes: great circle routes and distances were poorly presented, as distances in northern and southern latitudes were exaggerated.[17] More specifically, air power led to an enhanced demand for accurate maps, in order to plan and execute bombing and ground-support missions. Thus, in World War Two, maps of target areas were eagerly sought. Existing printed maps were acquired and supplemented by the products of photo-reconnaissance and other surveillance activity. The Germans used British Ordnance Survey maps as the basis for maps including information from photo-reconnaissance that they produced to guide their bombers during World War Two. Such photo-reconnaissance was also important for the creation of maps for amphibious and land operations, as with the Allied invasion of Normandy in 1944, or the German attack on the Soviet Union in 1941, which was preceded by long-range reconnaissance missions by high-flying Dornier Do-215 B2s and Heinkel He-111s.

Maps also played a role in the propaganda produced by the combatants in World War Two, serving to dramatize threats, as in posters showing Nazi Germany or Japan spreading their power. Frank Capra's film *Prelude to War*, produced by the Film Production Division of the US Army, used maps to underline the theme of challenge: the maps of Germany, Italy and Japan were transformed into menacing symbols, while the world map depicted the New World being surrounded and then conquered. Similar devices were employed by the Axis powers and their allies. There is, for example, a Vichy poster depicting Churchill as an octopus whose arms are reaching out for French possessions such as Syria and Dakar. Maps were also used to dramatize success, and thus to maintain civilian morale.

The more general problems of mapping war include not only data availability and accuracy, but also the problems of mapping what is by its very nature a dynamic subject. In addition, military mapping ignores the loss and suffering that is integral to war; it sanitizes the process of battle. Mapping is integral to war-gaming, whether by enthusiasts or armies, and translates the ethos of that bloodless practice to current conflicts.

Military mapping also emphasizes the fighting aspect of war, rather than other crucial dimensions, such as logistics and industrial capability. Furthermore, in strategic terms, mapping centres on a linear conception of war as a series of grand campaigns that offers little help to those who wish to understand 'popular' wars involving guerrilla activity.

There is also the problem of terminology. The definition of military units, such as army group, army corps, division, brigade, regiment and battalion, varies and has varied between countries. In addition, the size of

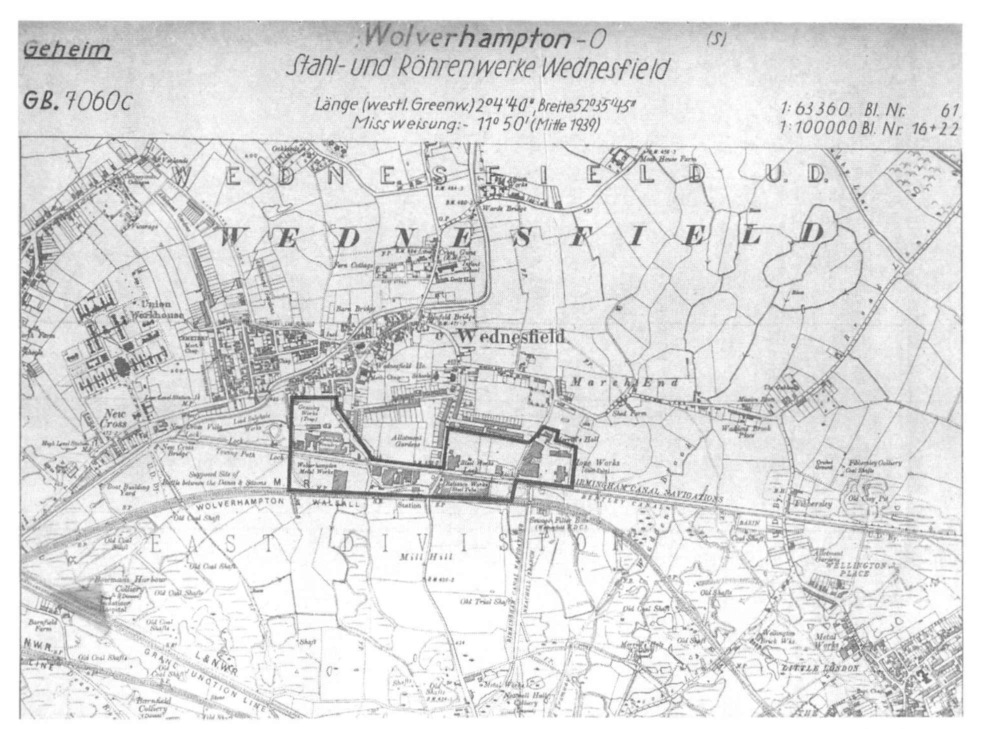

Luftwaffe map of Wolverhampton. Aerial reconnaissance supplemented existing printed maps in building up representations of target areas for bombing and ground-support operations.

such units varies and has varied, not least during the course of a single conflict. There are also problems at sea, as the designation of vessels has varied between countries.

It is impossible to achieve consistency. The different terms employed in titles, keys and accompanying text all have different connotations:

> A military occupation of hostile territory usually yields a mappable line of defense. If this line is temporary, military strategists and politicians might optimistically call it a *front*. If opposing forces have negotiated a cease-fire, it's termed a *truce line* or *armistice line*. A wide and somewhat ill-defined military line might be called a *zone*, whereas a longstanding, well-marked, and heavily fortified line is merely a *border*. When settlers as well as troops encroach on territory formerly or still partly inhabited by a weaker, less technologically advanced group, the line becomes a *frontier*.[18]

Such classifications and clarifications can be expanded endlessly in the search for a terminology as apparently precise, universal and objective as the lines on the map, but one man's 'atrocity' is another's 'pacification'. This is not simply a matter of debates within a single interpretative structure and tradition. The collapse of Soviet and eastern European

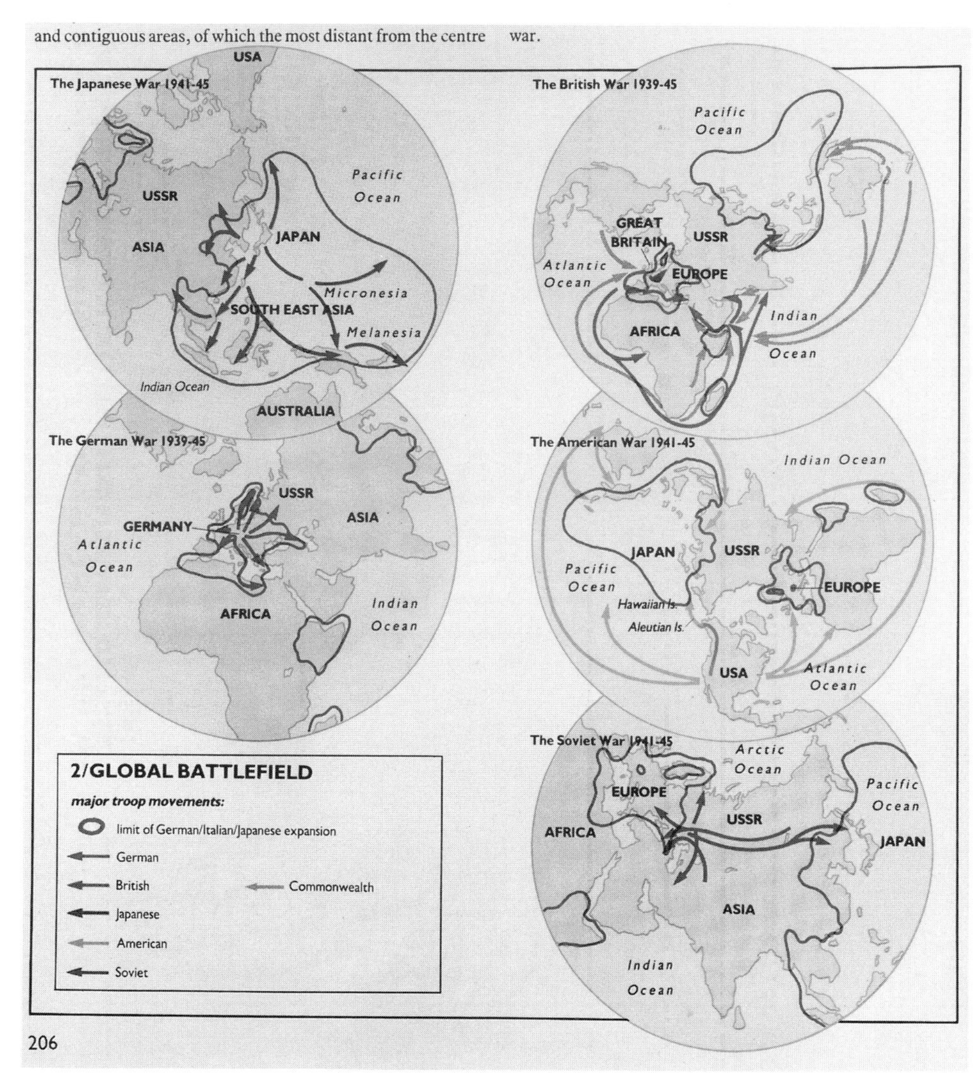

World War Two from different perspectives, from *The Times Atlas of the Second World War*. A valuable reminder of the extent to which different maps of the war can be presented by adopting varied perspectives. These maps create very different impressions.

Communism has challenged the interpretative strategies and descriptions of the Left, but the multiple discourses of nationalism compete over terms and terminology, and in both the 'tiger' economies of the Far East and the worlds of resurgent Islam there are challenges to the assessments dominant in the West. Where worlds intersect, as in the Arab–Israel struggle, there is no commonalty of understanding to serve as the basis for an agreed cartography. It is not simply that the location of lines on the map may be cause of controversy, but also that, and more importantly, the very significance of these lines is contested, and is contestable in a number of ways. This is true of both frontiers and wars: the two are related; they are different aspects of conflict. Thus what is aggressive in one light is defensive in another. This is readily apparent in maps. The ubiquitous advancing arrows of military maps can be presented in a more or less menacing fashion, and other aspects of cartographic symbolization can be utilized to the same ends.

Elections, frontiers and wars. These are the three major subjects that political maps are expected to show. All pose cartographic problems, but those of the last are greatest, because the dynamic element is integral to war. This is difficult to show. In addition, 'war is always a far worse muddle than anything you can produce in peace'.[19] Military events rarely conform to plans, and, in the height of advance, retreat and battle, there is generally considerable confusion about the location of units: units themselves amalgamate and fracture; command links are severed. War therefore lacks the precision of most election results. There are, of course, differences over modern frontiers, but each has a case law, and the differences are usually between competing precisely drawn lines.

The difficulties of mapping war have increased, as it has become more complex, on land, sea and air, and as it has expanded to encompass guerrilla operations more frequently. The conventional nature of military mapping centred on the concept of a front, a combination of units and positions that could be shown with some clarity and that advanced or retreated. Thus, lines and arrows summarized war. The situation at sea was not very different. Fleets advanced and retreated, and blockading positions could be expressed by lines. However, it was difficult to get across the time/space ratio on a map. Maps of Napoleonic campaigns, especially of 1805–7 and 1808–9, are of greater value if they can emphasize how quickly and efficiently Napoleon's armies moved across large distances and can contrast that to the sluggishness of their opponents.

Lines and arrows are less appropriate for modern war. They have been rendered less helpful by the range of modern weaponry and weapons systems. Aeroplanes and rockets overfly positions and can survey and destroy them in a way in which artillery could not. Gas can 'control' the atmosphere. The position of surface ships can be challenged by aeroplanes, missiles and submarines. In short, space in modern war is not controlled, certainly no longer controlled by troops or ships deployed in positions that can be readily depicted. Instead, space in war is multidimensional, and each of these dimensions is open to temporal variation, for example air sorties.

The notion of a front is now far less helpful than it was, for example, during the European wars of the mid-nineteenth century. Nevertheless, such a device is still the preferred one of newspaper maps. Their two-dimensional, black-and-white, character reduces the range and complexity of information that can be shown, and they are also affected by readers, who expect to see military mapping that is easy to understand.

Problems in depicting modern 'conventional' or 'regular' warfare (itself a construct) are matched by problems with 'irregular', 'unconventional' or 'guerrilla' warfare. Yet the latter has become more important, in part because of the politicization of the twentieth century, in particular in the 'Third World', a concept that is itself difficult to depict on maps. In 'irregular' warfare the notion of control over territory is challenged by forces that cannot be readily described in terms of conventional military units. They seek to operate from within the civilian population and do so not only for cover and sustenance, but also in order to deny their opponents any unchallenged control over populated areas. Guerrillas do not generally seek to gain control over regions, because that would provide their opponents with targets for their superior firepower. Instead, there is a system of shared presence; classically, one in which military or police patrols move unhindered, or suffer occasional sniping and ambushes and have to consider mines, but, otherwise, have no power: they control little beyond the ground they stand on. To map such a situation is extremely difficult. It can be mapped temporally – with the forces of authority shown as in control during the day, their opponents at night time – or spatially. The latter poses problems. Generally, the forces of authority operate along and seek to control communications routes, which are used for patrol and supply, while their presence in other areas is less common.

Yet the situation became more complicated with the development of aerial supply, as used, among others, by the French in Indo-China in the early 1950s, the United States in Vietnam, the British in Northern Ireland and the government of Sudan in southern Sudan. Aerial supply and operational capabilities were enhanced from the 1960s by effective and more powerful helicopters, for example Hueys in Vietnam, and American Cobras used by the Israelis in Lebanon. Vulnerable positions could be supplied and reinforced by air. Offensive operations could be mounted by helicopter, sometimes, as in Vietnam, in large-scale and sustained operations.

However, aerial resupply and attack capabilities were in turn limited by anti-aircraft weaponry, especially heat-seeking surface-to-air missiles. It is difficult to indicate their actual and potential impact on maps. The Israeli air force lost many planes to such missiles in the 1973 Yom Kippur war. The safety of low-level operations was therefore limited. The vertical space of the aerial battlefield was greatly affected. In the 1990–1 Gulf

War, mobile missile-launchers and mobile radars that 'locked' on to aeroplanes were both important, and, yet, again, it is difficult to map the role of such weapons, especially if the map is supposed to cover a range: either of time or weapons.

The problem is exacerbated if such weaponry is controlled by guerrilla forces. For example, by 1996, the IRA was supposed to have several Sam-7 surface-to-air missiles, while irregular forces in the southern Sudan use such missiles to challenge the resupply of garrisons by government planes. Thus, front lines dissolve. Just as the mapping of the political control of territory is made more difficult by theories and practices of shared sovereignty, and the cartographic device of frontiers is thus made to appear misleading, so the same is true of warfare. However, the added complexity of a more dynamic subject makes the mapping of war even more difficult.

When combined with guerrilla warfare the situation is further politicized. The terminology used is crucial to the legitimation or delegitimation of such warfare: guerrillas can be 'freedom fighters', 'bandits' or 'terrorists'; their actions 'war' or 'terrorism'; the response 'pacification' or 'atrocities'. Furthermore, whatever the terminology, very different messages are conveyed by the cartographic techniques used to convey guerrilla strength and operations, for example bombing campaigns. In addition, they can be shown to control or be operative in areas that vary greatly in size, depending on the political perspective of the map-maker. Moreover, the emphasis can be on their activities or on the response, or, alternatively, on the bulk of social activity that the guerrillas do not affect. Thus, for example, maps of the IRA or of Shining Path ('Sendero Luminoso') activities in Peru can suggest they are central or marginal; as earlier with maps of the Viet Cong. The very mapping of guerrilla activity can itself be a statement about the nature of a war. When, in 1958, the Bulgarians published an atlas of Bulgarian resistance in 1941–4, they were asserting that Bulgarian identity could be seen in anti-Fascist terms; that the presence of Bulgaria in the Eastern Bloc was not simply the consequence of Soviet conquest, but that that conquest in 1944 accorded with Bulgarian history.

More generally, the mapping of war, contemporary or historical, raises the issue of what should be shown. It is scarcely surprising that combatants provide cartographic accounts that stress their own role and interests at the expense of their allies and opponents (the two are not always clearly distinguished). Thus, British or American mapping of World War Two exaggerated their own contribution at the expense of that of the Soviets, and this is still the case.[20] It is also difficult to illustrate the psychological dimension of a conflict. In the postwar world, Berlin or Cuba took on symbolic importance far beyond any importance of location. Similarly, for many, Vietnam or Bosnia became ubiquitous presences.[21]

War, then, cannot be seen as an 'objective' sphere of mapping. The major problems involved in its cartographic analysis and depiction interact with subjective assessments of what is important, and there is also an important element of politicization, especially in the mapping of guerrilla warfare. These problems affect all aspects of the mapping of war, although they are generally neglected, especially in the public mapping of war, whether in newspapers or in war-gaming.[22] The use of maps in war-gaming has a long history, and was given a great boost by the spread of a self-consciously 'scientific' approach to military planning in the era of the general staffs that began in the late nineteenth century. At Oxford the University *Kriegspiel* [War-Games] Club 'played war games in the German style on the old set of Prussian official maps for the campaign of Sadowa [1866], and occasionally, for a change, on the Ordnance Survey maps of Oxfordshire, or the vicinity of Aldershot'. Its President was Hereford Brooke George (1838–1910), a pioneer of military history and geography at the university.[23] *Kriegspiel* had been introduced at the British Royal Military College in 1871.

Aside from such public depiction, mapping is also important as an adjunct not only of strategy but also of weaponry. 'Smart' weaponry, such as guided bombs and missiles, make use of the precise prior mapping of target and traverse in order to follow predetermined courses to targets that are actualized for the weapon as a grid reference. This was seen in the US cruise-missile attacks on Iraq in the Gulf War in 1991 and, subsequently, in 1995–6: the missiles used digital terrain models of the intended flight path. Other uses of modern mapping technology during the Gulf War included precise positioning devices interacting with US satellites in a Global Positioning System that was employed with success by Allied tanks in combat with those of Iraq, and the use of satellite imagery in the rapid production of photo-maps. In training exercises, modern armies use imaging GIS (Geographical Information System) software to provide instantaneous two- and three-dimensional views of battlefield exercise areas. The processes by which mapping affects conflict have become more insistent and rapid.

Mapping represents an important aspect of the information that is sought for the successful conduct of war. The gathering of intelligence about opponents has often centred and centres on the location of their forces, while, more generally, there was concern to establish an accurate account of the geography of opponents' territory. The acquisition of foreign maps, as for example by the British Intelligence Branch from the 1870s, played a major role, as did the compilation of new maps.

Intelligence was complemented by deception, which included the attempt to create false cartographies, to mislead opponents about the location and movements of forces. In World War Two, extensive and detailed deception schemes were designed to mislead the Germans about the Allied plans for a Second Front in Europe. These schemes involved false radio traffic, squadrons of wooden tanks, movements of landing craft and the creation of dummy landing craft, preparatory bombing raids that suggested the Allies would aim for Boulogne, fictional army movements and misinformation fed via agents. These varied means combined to suggest invasions directed against Boulogne, Brittany, the Balkans or Stavanger in Norway. A misleading cartography based on false intelligence was thus built up by the Germans. Similarly, the Russians were expert in major strategic deceptions. One such *maskirovka* in 1944 convinced the Germans that they were to be attacked south, not north, of the Pripet Marshes.

Intelligence gathering was and is a central aspect of the relationship between cartography and military concerns. It is not a relationship that has diminished with time. More generally, maps as a form, expression and organization of information are valuable to those who take decisions, and these are the wielders of power.

7 Conclusion

> Its very existence asserts the claim that Canada is a geographical entity. The atlas is shot through with the subtle geographic determinism which allowed Innis to insist that 'Canada' was co-extensive with the dependent fur-trading hinterland of northern North America whose political economy was sharply different from that of the agrarian colonies further south ... Certainly, there is a more marked emphasis upon the unities of Canadian history and the progression *towards* the unified nation state than might seem wholly plausible in the later 1990s. Demographic and cultural variety, the distinctiveness of Quebec, the strength of centripetal forces, the rights and claims of Native peoples (now termed the 'first nations') all receive somewhat less emphasis than might now be thought proper.[1]

The degree to which the selection, content and context of maps reflect assumptions is clear. It is also apparent that these assumptions can be regarded as political, in the wider sense of the word. The argument that knowledge is an adjunct, aspect, cause and consequence of power is also well established. Thus the loss of an understanding of maps and mapmaking as autonomous processes is but part of a wider challenge to conventional assessments of objectivity.

In the case of maps, there is also an additional understanding of the process of selection involved both in deciding what to map and how to map it, and in how to understand maps. Thus maps and assessments of them are affected by debates about texts and the role of writers and readers. They are also affected by notions of the visual and the role of iconography. Like caricatures, and unlike texts and pictures, maps are a fusion of literary and graphic, and both their meaning and their potency derive in part from this fusion. In the case of maps, but not caricatures, the two are in part separated. Attention is drawn to the map, not the key and the nomenclature of place names. There is no equivalent to the bubbles containing words coming from the mouths of those caricatures, with the exception of maps that use box-devices to a similar end. Yet, maps are similar to caricatures in some respects. The title, in particular, is a matter of choice; it is not obvious. To map, say, 'IRA terrorism' or 'Israeli aggression' or 'Poverty' is to make statements about what is being depicted that are in themselves contentious. The colours or shades defined in the key similarly arouse responses, although these are less conscious: some colours are more soothing and positive to the average viewer than others.

The analogy of maps and caricatures can be taken further. The purpose of advancing it is to link what is commonly seen as objective by intention, content and technique of production with what is commonly seen as subjective by all three criteria. Like caricatures, maps are political and politicizing texts that need to be read with care, although, unlike with caricatures, territoriality is a cartographic imperative and issue. Yet to problematize cartography is not to deny it value as a means of analysis or as a method of depiction. Indeed, maps are becoming more common and useful, both because of new techniques of map production and because of the perceived need to improve the 'packaging' of news items in a world where different media compete.[2] In societies that are more 'visual', or, rather, increasingly associate the authority of print with visual aspects and 'see the news' in visual terms, maps can and will play a greater role. Much will revolve around their traditional function of locating items: the map as plan, but the location can be of problems as much as places.

The authority, authentication and value of the map is an aspect of modern societies that are used to seeing plans that explain everything from car engines to knee joints and brains. Plans therefore reflect a society that both seeks to understand and that can create, construct and control. The two are increasingly part of the same process, indeed, mechanism. The notion of humanity as the observer that does not influence or control, the notion that underlay mapping of nineteenth-century geology or the plans of genetic inheritance of plant characteristics, is increasingly marginalized. Even maps of the cosmos, classically the mapping of something outside human control and understanding, now increasingly present a vision that has been grasped and comprehended and one that includes signs of human activity such as satellites and space debris. At the other extreme, Heffalumps do not exist outside of Winnie the Pooh's imagination, but he creates a trap for them, places it on his map, and the Heffalumps acquire spatial reality and cartographic presence. Pooh also finds the North Pole, 'appropriating' it with a notice proclaiming his discovery.

So the map as plan is the map as product and recorder of human agency, and, as such, affected by the controversial nature of such agency. The organic, unitary conception of human society that was so influential in the age of nineteenth-century nationalism appears less obvious at the close of the twentieth century. Democracy – the granting of the franchise – has been followed by democratization, in institutions, social concepts and practices, and knowledge. Grand strategies of understanding have fractured or been challenged.

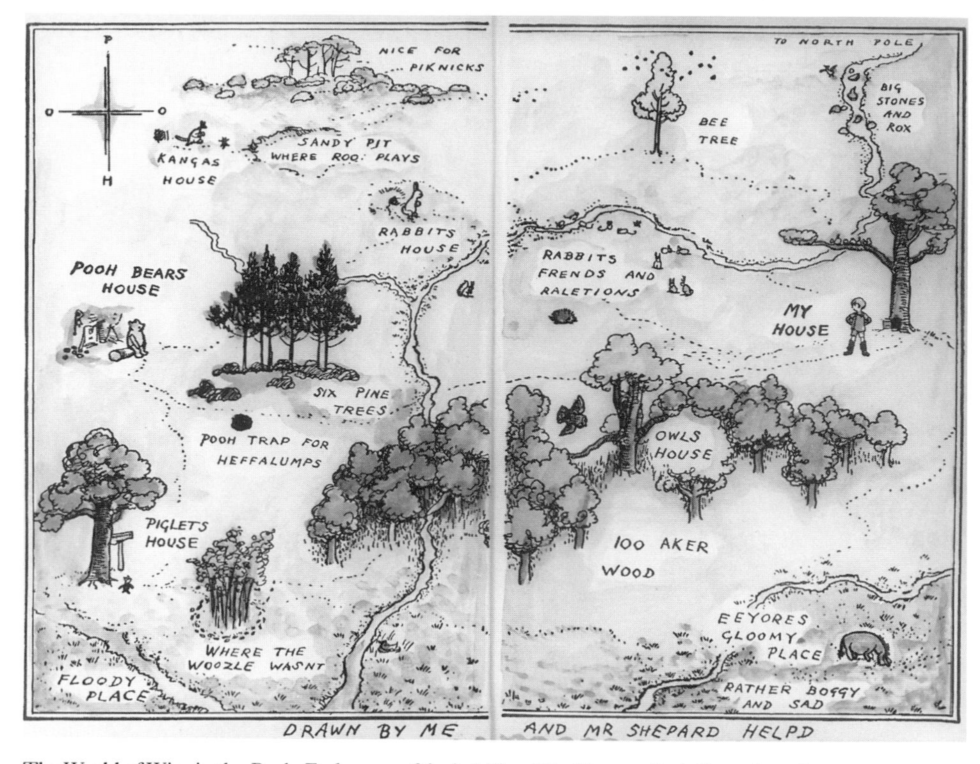

The World of Winnie-the-Pooh. Endpapers of A. A. Milne, *The House at Pooh Corner* (1974) edition).

These shifts affect the map as plan. What is to be shown becomes a matter for contention. Location itself is an issue. If, for example, the site of a confrontation on the West Bank is to be shown, certain questions arise: what boundaries should be employed; how should the West Bank and Israeli settlements be referred to; are Arab villages or Biblical sites to be shown; is reference to be made to the issue of water rights and, if so, how? Furthermore, does the map cover neighbouring Arab countries, so that the West Bank does not simply appear as an issue for Israel?

If the map is to be used as a tool for analysis the number of possible contentious issues increases, but, on the other hand, the very use of the map as a means for analysis, generally of a problem to which there are several apparent answers, makes it harder to treat the map as a simple objective statement separate to the process of human manipulation. Instead, the degree to which the map is used to indicate or assess a spatial relationship ensures that only apparently relevant aspects are mapped; the role of human choice, in establishing the experiment, is clear. If what is of relevance is itself controversial, then the controversy can be incorporated

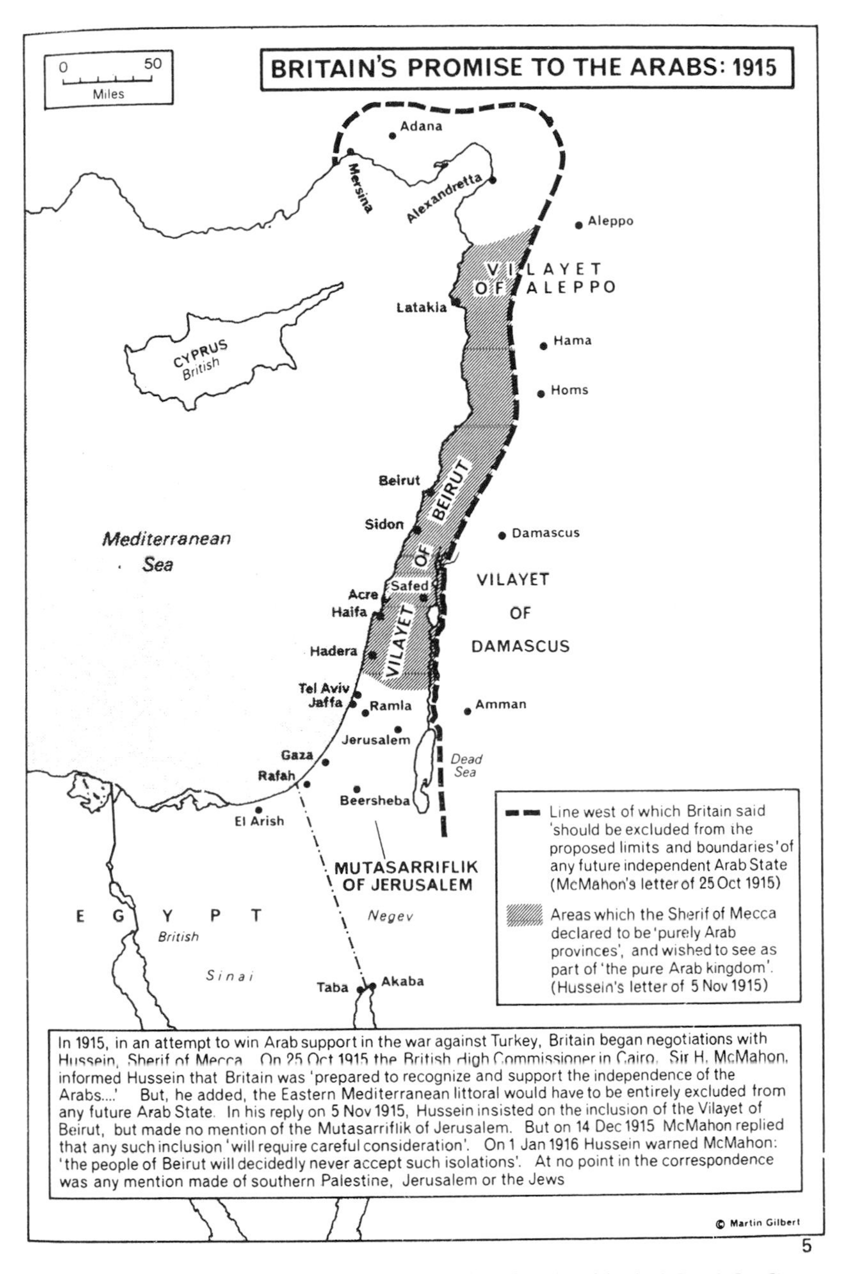

Britain's Promise to the Arabs, 1915, from Martin Gilbert's *Atlas of the Arab-Israeli Conflict* (1993).

into the process of map-making for analytical purposes by suggesting that the analysis should be refined and more, or different, factors mapped.

Mapping is therefore problematic. There is no unchallenged or obvious basis for an objectivity that can readily be used in order to challenge subjectivity or bias. Instead, subjectivity is central to the production and understanding of maps. Emphasis is introduced and assessment influenced by projection and perspective, colour and shading, scale and contouring, typography and key, order and combination. That should not, however, lead to an emphasis on maps as conspiratorial devices of the powerful. Some maps may indeed serve such ends, but maps are a medium not a message. The medium is multifaceted and, as with individual maps, it is possible to offer several analyses of its purpose and means of operation. An emphasis on subjectivity does not entail any suggestion that maps and map-making are without value, in fact, great value. Rather, it simply returns them to the social and political contexts in which they have meaning, not meaning without controversy, but meaning in controversy.

References

Introduction

1 T.P. Wiseman, 'Julius Caesar and the Mappa Mundi', in Wiseman, *Talking to Virgil: A Miscellany* (Exeter, 1992), pp. 33–40.
2 D.N. Livingstone, 'The Spaces of Knowledge: Contributions Towards a Historical Geography of Science', *Environment and Planning D: Society and Space*, 13 (1995), pp. 5–34.

1 *Cartography as Power*

1 J. Szegö, *Mapping Hidden Dimensions of the Urban Scene: Modelling the Cartographic Anatomy and Internal Dynamics of Growing Towns and Cities for Application in Urban and Regional Planning and in Environmental Analysis* (Stockholm, 1994).
2 J.B. Harley, 'Maps, Knowledge and Power', in D. Cosgrove and S. Daniels, eds., *The Iconography of Landscape* (Cambridge, 1988), pp. 277–312, 'Deconstructing the Map', in T. Barnes and J. Duncan, eds., *Writing Worlds: Discourse, Text and Metaphor in the Representation of the Landscape* (London, 1992), pp. 231–47, and 'Rereading the Maps of the Columbian Encounter', *Annals of the Association of American Geographers*, 82 (1992), pp. 522–42. For critiques of Harley, B. Belyea, 'Images of Power: Derrida/Foucault/Harley', *Cartographica*, 29/2 (1992), pp. 1–9, and J. H. Andrews, 'Meaning, Knowledge and Power in the Map Philosophy of J. B. Harley', *Trinity Papers in Geography*, 6 (1994). A positive and valuable review of his contribution that charts the developments in his ideas is provided by M.H. Edney, 'J. B. Harley (1932–1991): Questioning Maps, Questioning Cartography, Questioning Cartographers', *Cartography and Geographic Information Systems*, 19 (1992), pp. 175–8. Another important obituary was by W. Ravenhill in *Transactions of the Institute of British Geographers*, new series, 17 (1992), pp. 363–9.
3 J. Crampton, *Harley's Critical Cartography: In Search of a Language of Rhetoric* (University of Portsmouth, Department of Geography, Working Papers), 26 (1993).
4 I. C. Taylor, 'Official Geography and the Creation of "Canada"', *Cartographica*, 31/4 (Winter 1994), p. 1.
5 Harley, 'Silences and Secrecy: The Hidden Agenda of Cartography in Early Modern Europe', *Imago Mundi*, 40 (1988), pp. 58–9.
6 Harley, 'Cartography, Ethics and Social Theory', *Cartographica*, 27/2 (1990), pp. 4, 6. For Foucault, see F. Driver, 'Power, Space and the Body: A Critical Assessment of Foucault's *Discipline and Punishment*', *Environment and Planning D: Society and Space*, 3 (1985), pp. 425–46.
7 M. Pelletier, 'La Martinique et La Guadeloupe au lendemain du Traité de Paris (10 février 1763) l'oeuvre des ingénieurs géographes', *Chronique d'histoire maritime*, 9 (1984), pp. 22–30.
8 J. D. Forbes, *Atlas of Native History* (Davis, 1981), Introduction, unpaginated. See

also H. Brody, *Maps and Dreams: Indians and the British Columbia Frontier* (London, 1981); D. Turnbull, *Maps are Territories: Science is an Atlas* (Geelong, Victoria, 1989).

9 J. F. Ade Ajayi and M. Crowder, *Historical Atlas of Africa* (Harlow, 1985), Introduction, unpaginated.

10 D. Aberley, 'The Lure of Mapping: An Introduction', in Aberley, ed., *Boundaries of Home*, pp. 1–2.

11 M. H. Edney, 'Cartography without "Progress": Reinterpreting the Nature and Historical Development of Mapmaking', *Cartographica*, 30/23 (1993), pp. 54–68; S. S. Hall, *Mapping the Next Millennium. The Discovery of New Geographies* (New York, 1992), pp. 372, 383, 397–8; C. Jacob, *L'Empire des cartes: Approche théorique de la cartographie à travers l'histoire* (Paris, 1992), p. 457.

12 D. Hayden, *The Power of Place: Urban Landscapes as Public History* (Cambridge MA, 1995).

13 Taylor, 'Official Geography and the Creation of "Canada" ', *Cartographica*, 31/4 (Winter 1994), pp. 1–15.

14 D. Turnbull, 'Cartography and Science in Early Modern Europe: Mapping the Construction of Knowledge Spaces', *Imago Mundi*, 48 (1996), pp. 7, 19. See also E. Ferrier, 'Mapping Power: Cartography and Contemporary Cultural Theory', *Antithesis*, 1 (1990), p. 38.

15 A. M. MacEachren, *How Maps Work* (New York, 1995), p. 459; H. Foster (ed.), *Vision and Visuality* (Seattle, 1988); M. Jay, *Downcast Eyes* (Berkeley, 1993); D. Levin, *Modernity and the Hegemony of Vision* (Berkeley, 1993); D. Gregory, *Geographical Imaginations* (Oxford, 1993). Gregory also addressed the pattern/process problem inherent in any cartographic interpretation in his 'Social Geometry: Notes on the Recovery of Spatial Structure', in P. R. Gould and G. Olsson (eds.), *A Search for Common Ground* (London, 1982).

16 Edney, 'Mathematical Cosmography and the Social Ideology of British Cartography, 1780–1820', *Imago Mundi*, 46 (1994), pp. 112, 109.

17 D. Wood, 'P. D. A. Harvey and Medieval Mapmaking: An Essay Review', *Cartographica*, 31/3 (Autumn 1994), p. 58.

18 J. H. Andrews, *A Paper Landscape: The Ordnance Survey in Nineteenth Century Ireland* (Oxford, 1975), pp. 119–26, and 'Irish Placenames and the Ordnance Survey', *Cartographica*, 31/3 (Autumn 1994), pp. 60–1.

19 S. Kern, *The Culture of Time and Space, 1880–1918* (Cambridge, MA, 1983).

20 R. Dennis, *English Industrial Cities of the Nineteenth Century: A Social Geography* (Cambridge, 1984), p. 11.

21 Harley, 'Maps, Knowledge and Power', p. 279.

22 B. Crow and A. Thomas, *Third World Atlas* (Milton Keynes, 1983), p. 24.

23 H. Lefebvre, *The Production of Space* (Oxford, 1991).

24 D. Bell and G. Valentine, eds., *Mapping Desire: Geographies of Sexualities* (London, 1995).

25 For example, H. Brody, *Maps and Dreams: Indians and the British Columbia Frontier* (London, 1981).

26 F. Ormeling, 'New Forms, Concepts, and Structures for European National Atlases', *Cartographic Perspectives*, 20 (Winter 1995), p. 13.

27 C. Board, 'Things Maps Won't Show Us: Reflections on the Impact of Security Issues on Map Design', in K. Rybaczuk and M. Blakemore, eds., *Mapping the Nations*, International Cartographic Association Conference Proceedings (London, 1991), p. 137.

28 Harley, 'Deconstructing the Map', *Cartographica*, 26/2 (1989), p. 1.

29 M. Foucault, *Power/Knowledge: Selected Interviews and Other Writings* (New York, 1980), p. 131; A. Ophir and S. Schaffer, 'The Place of Knowledge', *Science in Context*, 4 (1991), pp. 3–21; F. Driver, 'Geography and Power: The Work of Michel Foucault', in P. Burke, ed., *Critical Essays on Michel Foucault* (Aldershot, 1992), pp. 147–56; C.

Philo, 'Foucault's Geography', *Environment and Planning D: Society and Space*, 10 (1992), pp. 137–61; Driver, 'Making Space', *Ecumene*, 1 (1994), pp. 386–90; D. Livingstone, 'The Spaces of Knowledge: Contributions Towards a Historical Geography of Science', *Environment and Planning D: Society and Space*, 13 (1995), pp. 5–34.

30 M. Warhus, *Another America: Native American Maps and the History of Our Land* (New York, 1997); N. Peterson, 'Totemism Yesterday: Sentiment and Local Organisation among the Australian Aborigines', *Man*, 7 (1972), pp. 12–32; Peterson and Langton, eds., *Aborigines, Land, and Land Rights* (Canberra, 1983); N. Williams, *The Yolgnu and their Land* (Canberra, 1986); H. Watson, 'Aboriginal-Australian Maps', in D. Turnbull, *Maps are Territories: Science is an Atlas* (Chicago, 1993), pp. 28–36; J. M. Jacobs, ' "Shake 'im This Country": The Mapping of the Aboriginal Sacred in Australia – The Case of Coronation Hill', in P. Jackson and J. Penrose, eds., *Constructions of Race, Place and Nation* (London, 1993), pp. 100–18.

31 A good example of a different spatiality is provided by the work of Denys Lombard, *Le Carrefour javanais: Essai d'histoire globale* (3 vols., Paris, 1990). On this see R. De Koninck, '*Le Carrefour javanais* de Denys Lombard', *Cahiers de géographie du Québec*, 36 (1992), pp. 339–45, and 'Au Carrefour de l'histoire et de la géographie: L'Île de Java selon Denys Lombard', *Mappemonde*, 4/92 (1992), pp. 42–4.

32 For example, B. J. Graham, 'No Place of the Mind: Contested Protestant Representations of Ulster', *Ecumene*, 1 (1994), pp. 257–81.

33 P. Nora, *Les Lieux de mémoire* (Paris, 1984–92); A. Charlesworth, 'Contesting Places of Memory: The Cause of Auschwitz', *Environment and Planning D: Society and Space*, 12 (1994), pp. 579–93.

34 Harley and Woodward, eds., *The History of Cartography* (Chicago, 1987–), *IIii. Cartography in the Traditional East and Southeast Asian Societies* (1994).

35 *Ibid*, Iii. *Cartography in the Traditional Islamic and South Asian Societies* (1992), p. 4.

2 *Mapping the World and its Peoples*

1 J. P. Snyder, *Flattening the Earth: Two Thousand Years of Map Projections* (Chicago, 1993), pp. 196–8.

2 *Ibid.*, pp. 258–62.

3 *Ibid.*, pp. 214–15.

4 A. Peters, *The New Cartography* (New York, 1983). See also W. L. Kaiser of the American National Council of Churches, 'New Global Map Presents Accurate Worldview', *Interracial Books for Children Bulletin*, 16 (1985), pp. 5–6, and *A New View of the World. A Handbook to the World Map: Peters Projection* (New York, 1987). The Peters and Kaiser books were both published by Friendship Press.

5 P. Vujakovic, 'The Extent of Adoption of the Peters Projection by "Third World" Organizations in the UK', *Society of University Cartographers Bulletin*, 21/1 (1987), pp. 11–15.

6 J. P. Snyder, 'Social Consciousness and World Maps', *The Christian Century*, 24 (February 1988), pp. 190–2.

7 D. H. Maling, 'Peters' Wunderwerk', *Kartographische Nachrichten*, 4 (1973), pp. 153–6. Despite its title, this was a critical article; J. Loxton, 'The Peters' Phenomenon', *The Cartographic Journal*, 22/2 (1985), pp. 106–8; A. H. Robinson, 'Arno Peters and His New Cartography', *The American Cartographer*, 12 (1985), pp. 103–11, and 'Reflections on the Gall-Peters Projection', *Social Education*, 51 (1987), pp. 260–4; P. Porter and P. Voxland, 'Distortion in Maps: The Peters' Projection and Other Devilments', *Focus*, 36 (1986), pp. 22–30; H. A. Sandford review of N. Myers, ed., *The Gaia Atlas of Planet Management* (London, 1985) in *Bulletin of the Society of University Cartographers*, 20/1 (1986), pp. 39–40; U. Freitag, 'Do We Need a New Cartography?', *Nachrichten aus dem Karten- und Vermassungswesen*, series 2/46 (1987),

pp. 51–9; Vujakovic, '*Peters Atlas*: A New Era of Cartography or Publisher's Contrick?', *Geography*, 74 (1989), pp. 245–51, and 'Arno Peters' Cult of the "New Cartography": From Concept to World Atlas', *Society of University Cartographers Bulletin*, 21/2 (1989), pp. 1–16.

8 J. Crampton, 'Cartography's Defining Moment: The Peters Projection Controversy', *Cartographica*, 31/4 (Winter 1994), pp. 16–32.

9 A. Spilhaus, 'Maps of the Whole World Ocean', *Geographical Review*, 32 (1942), pp. 431–5', 'To See the Oceans Slice Up the Land', *Smithsonian*, 10/8 (November 1979), p. 116, and 'World Ocean Maps: The Proper Places to Interrupt', *Proceedings of the American Philosophical Society*, 127/1 (January 1983), pp. 50–60; Spilhaus and J. P. Snyder, 'World Maps with Natural Boundaries', *Cartography and Geographic Information Systems*, 18 (1991), pp. 246–54.

10 M. H. Edney, 'Cartographic Confusion and Nationalism: The Washington Meridian in the Early Nineteenth Century', *Mapline*, 69–70 (1993), pp. 48.

11 A. K. Henrikson, 'America's Changing Place in the World: From "Periphery" to "Centre"?', in J. Gottman, ed., *Centre and Periphery: Spatial Variation in Politics* (Beverly Hills, 1980), pp. 79–80.

12 W. W. Ristow, *Maps for an Emerging Nation: Commercial Cartography in Nineteenth Century America* (Washington, 1977), pp. 32–3, 36; C. Gilbert, 'The End of Selective Availability', *Mapping Awareness*, 10/6 (July 1996), p. 10; A. Mason, *The Children's Atlas of Exploration* (London, 1993), pp. 8–9. See, more generally, K. Hodgkinson, 'Eurocentric World Views – The Hidden Curriculum of Humanities Maps and Atlases', *Multicultural Teaching*, 5/2 (1987), pp. 27–31, and 'Standing the World on its Head: A Review of Eurocentrism in Humanities Maps and Atlases', *Teaching History*, 62 (January 1991), pp. 19–23; R. A. Rundstrom, 'Mapping, Postmodernism, Indigenous People and the Changing Direction of North American Cartography', *Cartographica*, 28 (1991), pp. 112; C. A. Lutz and J. L . Collins, *Reading National Geographic* (Chicago, 1993), but see the critical review by Susan Schulten in *Reviews in American History*, 23 (1995), pp. 521–7.

13 J. Schwartzberg, ed., *Atlas of South Asian History* (2nd edn, New York, 1992), p. xxix.

14 R. Cooper, 'A Note on the Biological Concept of Race and its Application in Epidemiological Research', *American Medical Journal*, 108 (1984), pp. 715–23; P. A. Senior and R. Bhopal, 'Ethnicity as a Variable in Epidemiological Research', *British Medical Journal*, 309 (1994), pp. 327–30.

15 P. J. Stickler, 'Invisible Towns: A Case Study in the Cartography of South Africa', *Geographical Journal*, 22 (1990), pp. 329–33.

3 *Socio-Economic Issues and Cartography*

1 J. R. Akerman, 'From Books with Maps to Books as Maps: The Editor in the Creation of the Atlas Idea', in J. Winearls, ed., *Editing Early and Historical Atlases* (Toronto, 1995), pp. 3–48, esp. pp. 3–4. See also W.G. Dean, 'The Structure of Regional Atlases: An Essay on Communications', *Canadian Cartographer*, 7 (1970), pp. 48–60, and D. Wood, 'Pleasure in the Idea: The Atlas as Narrative Form', in R. J. B. Carswell, G. J. A. de Leeuw and N. M. Waters, eds., *Atlases for Schools: Design Principles and Curriculum Perspectives*, Cartographica monograph no. 36, *Cartographica*, 24/1 (1987), pp. 24–45.

2 D. Massey, 'The Geography of Trade Unions: Some Issues', *Transactions of the Institute of British Geographers*, new series 19 (1994), p. 98. See also Massey and J. Painter, 'The Changing Geography of Trade Unions', in J. Mohan, ed., *The Political Geography of Trade Unions* (Basingstoke, 1989), pp. 130–50.

3 D. Harvey, *The Limits to Capital* (Oxford, 1982); S. Corbridge, ed., *Money, Power and Space* (Oxford, 1994).

4 A. J. S. Gibson and T. C. Smout, 'Regional Prices and Market Regions: The

Evolution of the Early Modern Scottish Grain Market', *Economic History Review*, 48 (1995), p. 275; T. J. Barnes, *Logics of Dislocation: Models, Metaphors and Meanings of Economic Spaces* (Harlow, 1996).

5 H. Lefebvre, *The Production of Space* (Oxford, 1991).

6 D. Gregory, 'Historical Geography', in D. Gregory, P. Haggett, D. M. Smith and D. R. Stoddard, eds., *The Dictionary of Human Geography* (Oxford, 1981), pp. 146–50.

7 J. Hay, ed., *Boundaries in China* (London, 1994), pp. 12–13.

8 D. Hiebert, 'The Social Geography of Toronto in 1931: A Study of Residential Differentiation and Social Structure', *Journal of Historical Geography*, 21 (1995), p. 70.

9 B. Macdonald, *Vancouver. A Visual History* (Vancouver, 1992), pp. 70–1 for male–female proportions. D. Hayden, *The Grand Domestic Revolution: A History of Feminist Designs for American Homes, Neighbourhoods, and Cities* (Cambridge, MA, 1981).

10 For example, recently, H. L. Moore, *Space, Text and Gender: An Anthropological Study of the Marakwet of Kenya* (Harlow, 1996).

11 T. H. and C. C. Fast, *The Women's Atlas of the United States* (2nd edn, New York, 1995), pp. 1–2, 180–1.

12 M. P. Kumler and B. P. Buttenfield, 'Gender Differences in Map Reading Abilities: What Do We Know? What Can We Do?', in C. H. Wood and C. P. Keller, eds., *Cartographic Design: Theoretical and Practical Perspectives* (Chichester, 1996), pp. 125–36.

13 R. Downs and D. Stea, *Maps in Minds. Reflections on Cognitive Mapping* (New York, 1977), p. 239.

14 M. Lamont, 'National Identity and National Boundary Patterns in France and the United States', *French Historical Studies*, 19 (1995), p. 350; M. H. Matthews and P. Vujakovic, 'Private Worlds and Public Places: Mapping the Environmental Values of Wheelchair Users', *Environment and Planning A*, 27 (1995), pp. 1069–83; D. J. Bell, '[Screw]Ing Geography (Censor's Version)', *Environment and Planning D: Society and Space*, 13 (1995), pp. 127–31; B. Forest, 'West Hollywood as Symbol: The Significance of Place in the Construction of a Gay Identity', *ibid.*, pp. 133–57. See, more generally, P. Jackson, 'The Cultural Politics of Masculinity', *Transactions of the Institute of British Geographers*, 16 (1991), pp. 199–213; G. Rose, *Geography and Gender* (Oxford, 1993).

15 J. H. Mollenkopf, *New York City in the 1980s: A Social, Economic and Political Atlas* (New York, 1993), pp. 41–3; D. Rapetti, 'L'impôt dans la ville: de la rue aux quartiers nantais, 1972–80', *Mappe Monde*, 89/1 (1989), pp. 34–7.

16 M. Barke and R. J. Buswell, eds., *Newcastle's Changing Map* (Newcastle, 1992), p. 59.

17 C. A. Lutz and J. L. Collins, *Reading National Geographic* (Chicago, 1993); no author/editor, *The March of Civilization in Maps and Pictures* (New York, 1950), pt. III, p. 32; H. H. Kagan, ed., *The American Heritage Pictorial Atlas of United States History* (New York, 1966), p. 13.

18 E. A. Fernald and E. D. Purdum, eds., *Atlas of Florida* (Gainesville, FL, 1992), p. 81.

19 R. L. Bryant, 'Romancing Colonial Forestry: The Discourse of "Forestry as Progress" in British Burma', *Geographical Journal*, 162 (1996), pp. 169–78.

20 W. L. Kahrl, ed., *The California Water Atlas* (North Highlands, CA, 1978), pp. iv–vi, 3, 112.

21 M. Monmonier, *Drawing the Line, Tales of Maps and Cartocontroversy* (New York, 1995).

22 D. Aberley, ed., *Boundaries of Home: Mapping for Local Empowerment* (Philadelphia, PA, 1993).

23 S. Berthon and A. Robinson, *The Shape of the World* (London, 1991), p. 7.

24 I. C. Taylor, 'Official Geography and the Creation of "Canada" ', *Cartographica*, 31/4 (Winter 1994), pp. 12–13.

25 *Boundary and Security Bulletin*, 3/4 (Winter 1995–6), p. 12.

26 D. J. Dzurek, 'Eritrea–Yemen Dispute over the Hanish Islands', *Boundary and*

Security Bulletin, 4/1 (1996), pp. 70–7. Re the Spratlys, R. D. Hill, N. G. Owen and E.V. Roberts, eds., *Fishing in Troubled Waters* (Hong Kong, 1991). More generally, D. M. Johnston, *The Theory and History of Ocean Boundary Making* (Montréal, 1987).

27 Relevant recent works include J. Bulloch and A. Darwish, *Water Wars: Coming Conflicts in the Middle East* (London, 1993); P. P. Howell and J. A. Allan, *The Nile: Sharing a Scarce Resource: An Historical and Technical Review of Water Management and of Economic and Legal Issues* (Cambridge, 1994); N. Kliot, *Water Resources and Conflict in the Middle East* (London, 1994); J. A. Allan et al., *Water in the Middle East: Legal, Political and Commercial Implications* (London, 1995); A. T. Wolf, *Hydropolitics along the Jordan River: Scarce Water and its Impact on the Arab–Israeli Conflict* (New York, 1995).

28 M. Monmonier and G. A. Schnell, *Map Appreciation* (Englewood Cliffs, NJ, 1988), p. 245.

29 M. F. Goodchild, 'Stepping over the Line: Technological Constraints and the New Cartography', *The American Cartographer*, 15 (1988), pp. 311–19.

30 E. R. Tufte, *Envisioning Information* (Cheshire, CT, 1990).

31 D. Dorling, 'Cartograms for Visualizing Human Geography', in D. Unwin and H. Hearnshaw, eds., *Visualization and GIS* (Chichester, 1994), pp. 85–102, and 'Visualizing Changing Social Structure from a Census', *Environment and Planning A*, 27 (1995), pp. 353–78, and *A New Social Atlas of Britain* (Chichester, 1995).

32 J. Pickles, ed., *Grand Truth: The Social Implications of Geographic Information Systems* (New York, 1995).

33 J. Akerman, 'Selling Maps, Selling Highways: Rand McNally's "Blazed Trails" Program', *Imago Mundi*, 45 (1993), pp. 77–89, and 'Blazing a Well Worn Path: Cartographic Commercialism, Highway Promotion, and Auto Tourism in the United States, 1880–1930', *Cartographica*, 30/1 (Spring 1993), pp. 10–20.

34 W. W. Ristow, *Maps for An Emerging Nation: Commercial Cartography in Nineteenth-Century America* (Washington, DC, 1977), p. 27.

35 N. Nicholson, 'The First Family of American Maps', *Meridian*, 10 (1996), p. 8.

36 Richard Browne to his father, 24 August 1765, British Library, Department of Manuscripts, RP 3284.

37 R. V. Francaviglia, *The Shape of Texas: Maps and Metaphors* (College Station, TX, 1995). For the use of maps in advertising see also N. Holmes, ed., *Pictorial Maps* (London, 1992).

38 C. Scarre, 'The Western World View in Archaeological Atlases', in P. Gathercole and D. Lowenthal, eds., *The Politics of the Past* (London, 1990), pp. 11–18.

39 S. Rycroft and D. Cosgrove, 'Mapping the Modern Nation', *History Workshop Journal*, 40 (1995), pp. 91–105.

40 *The Times Atlas of the World* (London, 1968), xxviii; *The Reader's Digest Great World Atlas* (2nd edn, London, 1968), p. 119.

41 Z. Vilnay, *The New Israel Atlas: Bible to Present Day* (London, 1968), pp. 18, 29, 33–5.

4 *The Problems of Mapping Politics*

1 R. and B. Crampton, *Atlas of Eastern Europe in the Twentieth Century* (London, 1996), p. 156.

2 *Ibid.*, pp. 178–9.

3 J. B. Post, *An Atlas of Fantasy* (Baltimore, MD, 1973).

4 H. M. Hearnshaw and D. J. Unwin, eds., *Visualization in Geographic Information Systems* (Chichester, 1994); A. M. MacEachren, *Some Truth with Maps: A Primer on Symbolization and Design* (Washington, DC, 1994).

5 MacEachren, 'Visualising Uncertain Information', *Cartographic Perspectives*, 20 (1995), pp. 10–19.

6 MacEachren, *How Maps Work* (New York, 1995), pp. 276–7, 443.

7 K. Kox, 'The Voting Decision in Spatial Context', *Progress in Geography*, 1/2 (1969), pp. 96–100; K. Rohe, 'German Elections and Party Systems in Historical and Regional Perspective: An Introduction', in Rohe, ed., *Elections, Parties and Political Traditions* (New York, 1990), pp. 1–25; R. J. Johnston, *A Question of Place* (Oxford, 1991); P. Jehlicka, T. Kostelecky and L. Sykora, 'Czechoslovak Parliamentary Elections 1990: Old Patterns, New Trends and Lots of Surprises', in J. O'Loughlin and H. van der Wusten, eds., *The New Political Geography of Eastern Europe* (London, 1993), pp. 235–54; O'Loughlin, C. Flint and L. Anselin, 'The Geography of the Nazi Vote: Context, Confession, and Class in the Reichstag Election of 1930', *Annals of the Association of American Geographers*, 84 (1994), pp. 351–80; O'Loughlin, Flint and M. Shin, 'Regions and Milieux in Weimar Germany: The Nazi Party Vote of 1930 in Geographic Perspective', *Erdkunde*, 49 (1995), pp. 305–14; H. Carter, ed., *National Atlas of Wales* (1989), spread 2.2.

8 *Atlas des éléctions fédérales au Québec, 1867–1988* (1989).

9 G. Schoyer, 'The Coverage of Political Patterns and Elections in Some Selected State Atlases of the United States', *Special Libraries Association, Geography and Map Division, Bulletin*, 117 (September 1979), p. 7; J.V. Minghi, 'Politics', in J. F. Rooney, W. Zelinsky and D. R. Louder, eds., *This Remarkable Continent: An Atlas of United States and Canadian Society and Cultures* (College Station, TX), p. 207; M. Kinnear, *The British Voter. An Atlas and Survey since 1885* (London, 1981) is useful.

10 W. W. Ristow, *Maps for An Emerging Nation. Commercial Cartography in Nineteenth Century America* (Washington, DC, 1977), pp. 37, 63.

11 M. Ogborn, 'Local Power and State Regulation in Nineteenth Century Britain', *Transactions of the Institute of British Geographers*, n.s. 17 (1992), pp. 215–26.

12 M. D. Maltz, A. C. Gordon, W. Friedman, *Mapping Crime in its Community Setting. Event Geography Analysis* (New York, 1991). p. 21.

13 M. Vovelle, *La Decouverte de la politique: géopolitique de la Révolution Française* (Paris, 1993).

14 A. Dorpalen, *The World of General Haushofer: Geopolitics in Action* (New York, 1942).

15 J. M. Hunter, *Perspectives on Ratzel's Political Geography* (Lanham, MD, 1983), esp. pp. 252, 275; M. Bassim, 'Imperialism and the Nation-State in Fredrich Ratzel's Political Geography', *Progress in Human Geography*, 11 (1987), pp. 473–95; H. Mackinder, 'The Geographical Pivot of History', and subsequent discussion, *Geographical Journal*, 23 (1904), pp. 421–37, and *Democratic Ideals and Reality* (London, 1919). On Mackinder, W. H. Parker, *Mackinder: Geography as an Aid to Statecraft* (Oxford, 1982); B. W. Blouet, *Sir Halford Mackinder: A Biography* (College Station, TX, 1987); G. Ó. Tuathial, 'Putting Mackinder in his Place', *Political Geography*, 11 (1992), pp. 100–18. More generally, Parker, *Western Geopolitical Thought in the Twentieth Century* (London, 1985). For Amery, Mackinder, 'Geographical Pivot', p. 438.

16 A. K. Henrikson, 'Maps, Globes, and the "Cold War"', *Special Libraries*, 65 (1974), pp. 445–54.

17 E. L. Ayers, P. W. Limerick, S. Nissenbaum and P. S. Onuf, *All Over the Map. Rethinking American Regions* (Baltimore, MD, 1996), p. vii.

18 J. W. Konvitz, 'The Nation-State, Paris and Cartography in Eighteenth- and Nineteenth-Century France', *Journal of Historical Geography*, 16 (1990), pp. 37.

19 *National Atlas of Mongolia* (Moscow, 1990), pp. 20–2.

20 S. Schulten, 'The Transformation of World Geography in American Life, 1880–1950', unpublished abstract. I would like to thank Susan Schulten for sending me this abstract and other unpublished items.

21 Ex. inf. R. H. Hewsen.

22 M. Brawer, *Atlas of Russia and the Independent Republics* (New York, 1994), p. 6.

23 M. Monmonier and G. A. Schnell, *Map Appreciation* (Englewood Cliffs, NJ, 1988), p. 205.
24 S. Schulten, unpublished paper, 'Locating the World: Popular Cartography in the United States, 1880–1950'.
25 J. Elliott and C. King, *Usborne Children's Encyclopedia* (London, 1986), pp. 30–1, 26–7.
26 B. Williams, *Kingfisher First Encyclopedia* (London, 1994), pp. 9, 24–5.
27 H. R. Roemer, 'The Safavid Period', in P. Jackson and L. Lockhart, eds., *The Cambridge History of Iran VI* (Cambridge, 1986), p. 258.
28 M. Bassin, 'Expansion and Colonialism on the Eastern Frontier: Views of Siberia and the Far East in Pre-Petrine Russia', *Journal of Historical Geography*, 14 (1988), p. 16.
29 I. G. Taylor, 'Official Geography and the Creation of "Canada" ', *Cartographica*, 31/4 (Winter 1994), p. 14.
30 J. R. Akerman, 'The Structuring of Political Territory in Early Printed Atlases', *Imago Mundi*, 47 (1995), pp. 138–54.
31 J. Gottmann, *The Significance of Territory* (Charlottesville, VI, 1973).
32 H. S. Shapiro, 'Giving a Graphic Example: the Increasing Use of Charts and Maps', *Nieman Reports*, 36 (1982), pp. 4–7.
33 P. Gilmartin, 'The Design of Journalistic Maps/Purposes, Parameters and Prospects', *Cartographica*, 22 (1985), pp. 1–18.
34 *Sunday Times*, 17 November 1996, p. 9.
35 G. Ó Tuathial, 'Geopolitics and Discourse: Practical Geopolitical Reasoning in American Foreign Policy', *Political Geography*, 11 (1992), pp. 190–204; D. Slater, 'The Geopolitical Imagination and the Enframing of Development Theory', *Transactions of the Institute of British Geographers*, n.s. 18 (1993), pp. 419–37,and subsequent debate, 19 (1994), pp. 228–38; G. Ó. Tuathial and S. Dalby, editorial introduction to *Environment and Planning D: Society and Space* 12 (1994), pp. 513–634 issue on geopolitics, at p. 513; G. Ó. Tuathial, 'Problematising Geopolitics: Survey, Statesmanship and Strategy', *Transactions of the Institute of British Geographers*, 19 (1994), pp. 259–72, and *Critical Geopolitics: The Politics of Writing Global Space* (Minneapolis, 1996).

5 *Frontiers*

1 F. de Dainville, 'Cartes et contestations au XVe siècle', *Imago Mundi*, 24 (1970), pp. 99–121, for example p. 112 and figure 10; R. Almagia, *Monumenta italiae cartographica* (1929), p. 13, plate XIII.
2 S. R. Gammon, *Statesman and Schemer. William, First Lord Paget. Tudor Minister* (Newton Abbot, 1973), p. 106. On the early-modern period see, more generally, P. Barber, 'Maps and Monarchs in Europe 1550–1800', in R. Oresko, G. C. Gibbs and H. M. Scott, eds., *Royal and Republican Sovereignty in Early Modern Europe* (Cambridge, 1996), pp. 75–124; K. Buczek, *The History of Polish Cartography from the 15th to the 18th Century* (Amsterdam, 1982).
3 J. Richard, 'Enclaves royales et limites des provinces', *Annales de Bourgogne*, 20 (1948), p. 112.
4 C. J. Ekberg, *The Failure of Louis XIV's Dutch War* (Chapel Hill, 1979), pp. 118–19; P. Sahlins, 'Natural Frontiers Revisited: France's Boundaries since the Seventeenth Century', *American Historical Review*, 95 (1990), pp. 1, 433–4.
5 Louis XV to Vaulgrenant, 21 October 1745, Paris, Bibliothèque Victor Cousin, Fonds de Richelieu 40, f. 77; Puysieulx to Duke of Richelieu, French commander at Genoa, 22 July 1748, Paris, Archives Nationales, KK 1372.
6 P. Harsin, *Les Relations extérieures de la Principauté de Liège* (1927), p. 164; N. G.

d'Albissin, 'Propos sur la frontière', *Revue historique de droit Français et étranger*, 47 (1966), pp. 390–407.

7 Sahlins, *Boundaries: The Making of France and Spain in the Pyrenees* (Berkeley, 1989), pp. 35, 49, 187; Rebenac, French envoy in Madrid to Louis XIV, 9 September 1688, Paris, Archives du Ministère des Affaires Étrangères, Correspondance Politique (hereafter AE. CP.) Espagne 75, f. 56; Lord Grantham, British envoy in Madrid, to Horace St. Paul, Secretary of Embassy at Paris, 8 May, 20 November 1775, London, British Library, Additional Manuscripts (hereafter BL. Add.) 24177, ff. 41, 501.

8 Dainville, 'Cartes et contestations', p. 118.

9 P. Hetherington, 'Anglo-Scottish Borders', *Boundary Bulletin*, 2 (1991).

10 P. Waeber, *La Formation du Canton de Genève* (Geneva, 1974), pp. 42–3.

11 Praslin, French foreign minister, to Châtelet, French envoy in Vienna, 16 July 1763, AE. CP. Autriche 295, f. 54; P. de Lapradelle, *La Frontière: étude de droit international* (Paris, 1928), p. 45 n. 1; J. F. Noel, 'Les problèmes des frontières entre la France et l'Empire dans la seconde moitié du XVIIIe siècle', *Revue historique*, 235 (1946), pp. 336–7; G. Livet, ed., *Recueil des Instructions données aux Ambassadeurs et Ministres de France depuis les Traités de Westphalie jusqu'à la Révolution Française. L'Electorat de Trèves* (Paris, 1966), pp. cxix–cxx, cxxxiii–cxl; O.T. Murphy, *Charles Gravier: Comte de Vergennes* (Albany, 1982), p. 454; D. Nordman and J. Revel, 'La formation de l'espace français', in J. Revel, ed., *Histoire de la France, vol I: L'Espace français* (Paris, 1989), pp. 29–69; Sahlins, 'Natural Frontiers Revisited', pp. 1, 438–41.

12 A. Somme, ed., *A Geography of Norden* (London, 1968), pp. 15–17; D. Kirby, *Northern Europe in the Early Modern Period: The Baltic World 1492–1772* (Harlow, 1990), p. 20.

13 Keith to Lord Grenville, Foreign Secretary, 5 August 1791, London, Public Record Office, Foreign Office, 7/27, f. 167; *Gentleman's Magazine* (London, 1791), p. 861; K. A. Roider, *Austria's Eastern Question 1700–1790* (Princeton, NJ, 1982), pp. 177, 189.

14 H. Inalcik, 'Ottoman Methods of Conquest', *Studia Islamica*, 2 (1954), pp. 103–29.

15 O. Subtelny, *Domination of Eastern Europe: Native Nobilities and Foreign Absolutism, 1500–1715* (Gloucester, 1986).

16 V. W. Crane, *The Southern Frontier, 1670–1732* (Ann Arbor, MI, 1929); W.P. Cumming, *The Southeast in Early Maps* (Chapel Hill, NC, 1962); L. DeVorsey Jr., *The Indian Boundary in the Southern Colonies, 1763–1775* (Chapel Hill, NC, 1966); D. H. Cockran, *The Creek Frontier, 1540–1783* (Norman, OK, 1967); J. M. Sosin, *The Revolutionary Frontier, 1763–1783* (New York, 1967); E. J. Cashin, *Lachlan McGillivray, Indian Trader: The Shaping of the Southern Colonial Frontier* (Athens, GA, 1992), pp. 214–22, 229, 238–47.

17 R. A. Abou-El-Haj, 'The Formal Closure of the Ottoman Frontier in Europe: 1699–1703', *Journal of the American Oriental Society*, 89 (1969); J. Stoye, *The Life and Times of Luigi Ferdinando Marsigli, Soldier and Virtuoso* (New Haven, CT, 1993).

18 D. M. Lang, *The Last Years of the Georgian Monarchy 1658–1832* (New York, 1957); A. W. Fisher, *The Russian Annexation of the Crimea, 1772–1783* (Cambridge, 1970); G. Jewsbury, *The Russian Annexation of Bessarabia: 1774–1828* (Boulder, CO, 1976); M. Atkin, *Russia and Iran, 1780–1828* (Minneapolis, 1980).

19 D. K. Bassett, *British Trade and Policy in Indonesia and Malaysia in the Late Eighteenth Century* (Hull, 1971), pp. 73–80; R. Bonney, *Kedah 1771–1821: The Search for Security and Independence* (Oxford, 1971), pp. 52–101; C. A. Bayly, *Imperial Meridian: The British Empire and the World 1780–1830* (London, 1989), pp. 46–7.

20 J. D. Black, ed., *The Blathwayt Atlas* (2 vols, Providence, RI, 1970–5), vol. I, 49–55; P. Barber, 'Necessary and Ornamental: Map Use in England under the Later Stuarts, 1660–1714', *Eighteenth-Century Life*, 14 (1990), p. 19.

21 Joseph Yorke to Lord Chancellor Hardwicke, 27 August 1749, BL. Add. 35355, f. 103.

22 Mirepoix to Rouillé, French foreign minister, 16 January, 8 March 1755, AE. CP. Ang. 438 f. 18, 261.

23 Bonnac to Rouillé, 21 February 1755, AE. CP. Hollande 488 ff. 106–7; Bussy to Rouillé, 29 July 1755, AE. CP. Brunswick–Hanover 52, f. 22; A. Reese, *Europäische Hegemonie und France d'outre-mer. Koloniale Fragen in der französischen Aussenpolitik 1700–1763* (Stuttgart, 1988), pp. 274–310.
24 E. A. Reitan, 'Expanding Horizons: Maps in the *Gentleman's Magazine*, 1731–1754', *Imago Mundi*, 37 (1985), pp. 54–62.
25 Z. E. Rashed, *The Peace of Paris 1763* (Liverpool, 1951), p. 166 and map opposite p. 254.
26 J. L. Wright, *Britain and the American Frontier, 1783–1815* (Athens, GA, 1975); R. C. Stuart, *United States Expansionism and British North America, 1775–1871* (Chapel Hill, NC, 1988).
27 A. P. Whitaker, *The Spanish–American Frontier, 1783–1795* (Boston, 1927); S. F. Bemis, *Pinckney's Treaty: America's Advantage from Europe's Distress* (2nd edn, New Haven, CT, 1960).
28 R. C. Downes, *Evolution of Ohio County Boundaries* (Columbus, OH, 1970); *The Atlas of Pennsylvania* (Philadelphia, PA,1989), p. 81.
29 A. Godlewska and N. Smith, eds., *Geography and Empire* (Oxford, 1994).
30 D. Hooson, ed., *Geography and National Identity* (Oxford, 1994).
31 J. C. Stone, *A Short History of the Cartography of Africa* (Lewiston, 1995), pp. 107, 228–9; Sahlins, 'Centring the Periphery: The Cerdanya between France and Spain', in R. L. Kagan and G. Parker, eds., *Spain, Europe and the Atlantic World* (Cambridge, 1995), p. 231.
32 S. Berthon and A. Robinson, *The Shape of the World* (London, 1991), p. 169.
33 J. J. Ferguson and E. R. Hart, *A Zuni Atlas* (Norman, 1985), pp. 56–7.
34 M. S. Seligmann, 'Maps as the Progenitors of Territorial Disputes: Two Examples from Nineteenth-Century Southern Africa', *Imago Mundi*, 47 (1995), pp. 173–83; J. B. Harley, 'Maps, Knowledge, and Power', in D. Cosgrove and S. Daniels, eds., *The Iconography of Landscape* (Cambridge, 1988), p. 282; G. Huggan, 'Decolonising the Map: Postcolonialism, Poststructuralism and the Cartographic Connection', in I. Adam and H. Tiffin, eds., *Past the Last Post: Theorizing Postcolonialism and Postmodernism* (London, 1991); J.K. Noyes, *Colonial Space: Spatiality in the Discourse of German South West Africa, 1884–1915* (Chur, 1992); T. J. Bassett, 'Cartography and Empire Building in Nineteenth-Century West Africa', *Geographical Review*, 84 (1994), pp. 316–35; F. Driver, 'Geography's Empire: Histories of Geographical Knowledge', *Environment and Planing D: Society and Space*, 10 (1992), pp. 23–40; A. Godlewska and N. Smith, eds., *Geography and Europe* (Oxford, 1994).
35 R. Schofield, ed., *The Iran–Iraq Border 1840–1958* (11 vols., Neuchâtel, 1989).
36 Stone, *Cartography of Africa*, p. 228; Schofield and G. Blake, ed., *Arabian Boundaries: Primary Documents 1853–1960)* (30 vols., Slough, 1988).
37 R. Hay, 'The Persian Gulf States and their Boundary Problems', *Geographical Journal*, 120 (1954), p. 431.
38 R. Schofield, ed., *Islands and Maritime Boundaries of the Gulf 1798–1960* (20 vols., Neuchâtel, 1991).
39 J. C. Wilkinson, *Arabia's Frontiers: The Story of Britain's Boundary Drawing in the Desert* (London, 1991).
40 P. Toye, *Palestine Boundaries 1833–1947* (4 vols., Neuchâtel, 1989); D. Gavish, 'The British Efforts at Safeguarding the Land Records of Palestine in 1948', *Archives*, 22 (1996), pp. 107–20; M. Gilbert, *The Dent Atlas of the Arab–Israeli Conflict* (6th edn, London, 1993), p. 5.
41 F. W. Mote and D. Twitchett, eds., *The Cambridge History of China. VII: The Ming Dynasty, 1368–1644, Part 1* (Cambridge, 1988), pp. 392–3.
42 General Assembly Resolution 1514 of 14 December 1960.
43 S. J. Anaya, 'The Capacity of International Law to Advance Ethnic or Nationality Rights Claims', *Human Rights Quarterly*, 13 (1991), pp. 403–11, and J. J. Corntassel

and T. H. Primeau, 'Indigenous "Sovereignty" and International Law: Revised Strategies for Pursuing "Self-Determination"', *Human Rights Quarterly*, 17 (1995), pp. 140–56.

44 R.Griggs and P. Hocknell, 'Fourth World Faultlines and the Remaking of "International" Boundaries', International Boundaries Research Unit, *Boundary and Security Bulletin*, 3/3 (1995), pp. 49–58.

45 T. Winichakul, *Siam Mapped: a History of the Geo-Body of a Nation* (Honolulu, HI, 1994).

46 W. S. Miles, *Hausaland Divided: Colonialism and Independence in Nigeria and Niger* (Ithaca, NY, 1994). See also A. I. Asiwaju, ed., *Partitioned Africans: Ethnic Relations Across Africa's International Boundaries, 1884–1984* (London, 1985), and Asiwaju and P. Nugent, eds., *African Boundaries: Barriers, Conduits and Opportunities* (London, 1996).

47 M. Klinge, 'The Baltic An Image', in U. Ehrensvärd, P. Kokkonen and J. Nurminen, eds., *Mare Balticum* (Helsinki, 1995), p. 10.

48 E. Fredrickson, introduction, *Finland 500 Years on the Map of Europe* (Jyväskylä, Finland, 1993), p. 2.

49 For example, L. Boban, *Croatian Borders 1918–1993*, English translation of second edition of a Croatian work (Zagreb, 1993); *A Concise Atlas of the Republic of Croatia and of the Republic of Bosnia and Hercegovina* (Zagreb, 1993).

50 D. Rumley and J. V. Minghi, eds., *The Geography of Border Landscapes* (London, 1991).

6 *War as an Aspect of Political Cartography*

1 D. Buisseret, ed., *Monarchs, Ministers and Maps: The Emergence of Cartography as a Tool of Government in Early Modern Europe* (Chicago, 1992).

2 L. R. Shelby, *John Rogers: Tudor Military Engineer* (Oxford, 1967); R. A. Skelton, 'The Military Surveyor's Contribution to British Cartography in the Sixteenth Century', *Imago Mundi*, 24 (1970), pp. 77–83; D. W. Marshall, 'The British Military Engineers 1741–1783: A Study of Organization, Social Origin and Cartography' (unpublished PhD, Michigan, 1976); D. Hodson, *Maps of Portsmouth before 1801* (Portsmouth, 1978); W. A. Seymour, *A History of the Ordnance Survey* (Folkestone, 1980); E. Stuart, *Lost Landscapes of Plymouth: Maps, Charts and Plans to 1800* (Stroud, 1991); Harley and Woodward, *Cartography in the Traditional Islamic and South Asian Societies*, pp. 209–15, 462–5, 491–3.

3 U. Freitag, *Kartographische Konzeptionen/Cartographic Conceptions* (Berlin, 1992), p. 280.

4 *Literary Magazine*, 15 October 1756.

5 G. Raudzens, 'The British Ordnance Department and the Advancement of Geographic Science', *Cartography*, 1 (1991), pp. 106–9; J.W. Fireman, *The Spanish Corps of Engineers in the Western Borderlands: Instruments of Bourbon Reform 1764–1815* (Glendale, CA, 1977); M. Watelet, 'La cartographie topographique militaire des Alliés en France et en Belgique (1815–1818)', *Bulletin trimestriel du crédit communal de Belgique*, 174 (1990).

6 W. H. Goetzmann, *Army Exploration in the American West, 1803–63* (New Haven, CT, 1959); F. Schubert, *Vanguard of Expansion: Army Engineers in the Trans-Mississippi West, 1819–1879* (Washington, DC, 1980); A. G. Traas, *From The Golden Gate to Mexico City: The US Army Topographical Engineers in the Mexican War, 1846–1848* (Washington, DC, 1993).

7 M. Warhus, *Cartographic Encounters: An Exhibition of Native American Maps from Central Mexico to the Arctic, Mapline*, special issue, number 7, September 1993, pp. 15-16; F.C. Luebke, F. W. Kaye and G. E. Moulton, eds., *Mapping the North American Plains* (Norman, OK, 1987).

8 C. Nelson, *Mapping the Civil War* (Washington, DC, 1992); W. J. Miller, *Mapping for Stonewall: The Civil War Service of Jed Hotchkiss* (Washington, DC, 1993).
9 W. W. Ristow, *Maps for an Emerging Nation: Commercial Cartography in Nineteenth Century America* (Washington, DC, 1977), p. 29.
10 D. Bosse, *Civil War Newspaper Maps of the Northern Daily Press: A Cartobibliography* (Westport, CT, 1993) and *Civil War Newspaper Maps* (Baltimore, MD, 1993).
11 G. L. H. Davies, *North from the Hook: 150 Years of the Geological Survey of Ireland* (Dublin, 1995), pp. 299-300.
12 J. S. Murray, 'The Face of Armageddon', *Mercator's World*, 1/2 (1996), pp. 30–7; M. Heffernan, 'Geography, Cartography and Military Intelligence: The Royal Geographical Society and the First World War', *Transactions of the Institute of British Geographers*, 21 (1996), pp. 504–33.
13 P. McMaster, 'Ordnance Survey: Completing Two Centuries of National Mapping and Now Facing the Challenge of the 1990s', in K. Rybaczuk and M. Blakemore, eds., *Mapping the Nations* (International Cartographic Association Conference Proceedings), (2 vols., cont. pag., London, 1991), p. 7.
14 J. K. Wright, *Geography in the Making: The American Geographical Society 1851–1951* (New York, 1952), pp. 354-5; W. C. V. Balchin, 'United Kingdom Geographers in the Second World War', *Geographical Journal*, 153 (1987), pp. 159–80.
15 A. C. Hudson, 'The New York Public Library's Map Division Goes to War, 1941–1945', *Geography and Map Division: Special Libraries Association Bulletin*, 182 (Spring 1996), pp. 2–25.
16 Ristow, 'Journalistic Cartography', *Surveying and Mapping*, 17 (1957), pp. 369-90; M. Mandell, 'World War II Maps for Armchair Generals', *Mercator's World*, 1/4 (1996), pp. 42-5.
17 J. Ager, 'Maps and Propaganda', *Bulletin of the Society of University Cartographers*, 11/1 (1977), p. 8; A. K. Henrickson, 'The Map as "Idea": the Role of Cartographic Imagery during the Second World War', *The American Cartographer*, 2 (1975), pp. 19-53.
18 M. Monmonier, *Mapping It Out: Expository Cartography for the Humanities and Social Sciences* (Chicago, 1993), p. 200.
19 Field Marshal Wavell quoted in R. O'Shea and D. Greenspan, *American Heritage Battle Maps of the American Civil War* (Stroud, 1994), p. 7.
20 M. Kitchen, 'Old Tales of the Second World War', *International History Review*, 13 (1991), p. 110, with reference to B. and F. Pitt, *The Chronological Atlas of World War II* (London, 1989).
21 G. Ó. Tuathial, 'An Anti-geopolitical Eye: Maggie O'Kane in Bosnia', *Gender, Place and Culture*, 3 (1996), pp. 171–85.
22 The range of the latter is illustrated in W. W. Easton, 'War Games and Maps', *Geography and Map Division: Special Libraries Association Bulletin*, 111 (March 1978), pp. 18–24.
23 R. A. Butlin, 'Historical Geographies of the British Empire, c. 1887–1925', in M. Bell, R. A. Butlin and M. Heffernan, eds., *Geography and Imperialism 1820–1940* (Manchester, 1995), pp. 169–70.

7 Conclusion

1 J. G. Darwin, review of *Historical Atlas of Canada*, *Economic History Review*, 49 (1996), p. 416.
2 M. Monmonier, *Maps with the News* (Chicago, 1989), pp. 68, 244.

Picture Acknowledgements

The author and publishers wish to express their thanks to the following sources of illustrative material and/or permission to reproduce it:

© Akademische Verlagsanstalt 1997, for Peters Map: page 35; American Philosophical Society: page 38; M. Barke & R.J. Buswell: page 76; British Library, London: pages 131, 135, 149; Richard and Ben Crampton and Routledge, for *Atlas of Eastern Europe in the Twentieth Century*: page 101; Daniel Dorling: page 100; Librairie Arthème Fayard, Paris: page 83; for *First Encyclopedia* by Brian Williams, published by Kingfisher, © Grisewood & Dempsey Ltd 1991: page 117; for *The Gaia Atlas of Planet Management*, Gaia Books Ltd.: pages 50 (bottom) and 81 (artwork by Eugene Fleury) and page 80 (artwork by Chris Forsey); M. Gilbert and Routledge, for *Atlas of the Arab-Israeli Conflict*: page 167; HarperCollins Publishers: page 103, and © extracted from the Nicholson London Streetfinder Atlas, 1996, pages 14, 15; Crown copyright 87799M: pages 14, 15, 16; Kort & Matrikelstyrelsen, Copenhagen: page 91; London Transport Museum: page 49; Stuart McArthur/Artarom: pages 52–53; Macmillan Library Reference USA, a Simon and Schuster Macmillan Company, for *New York City in the 1980s: A Social, Economic, and Political Atlas*, by John H. Mollenkopf, copyright © 1993 by John H. Mollenkopf: page 51; © Michelin, for map no. 230 (22nd edition, 1997) – authorization number 9705281: page 56; © Myriad Editions, Inc., 1995, reproduced by permission of Frederick Warne & Co.: pages 50 (top), 54 (top), 70–72; National Archives of Canada: page 153; Quarto Children's Books: page 45; Reed Consumer Books: page 6; Royal Geographical Society, London: pages 34 (bottom), 111; for map illustration by E. H. Shepard, copyright under the Berne Convention, reproduced by permission of Curtis Brown, London: page 166; Times Books, reproduced with permission of HarperCollins (MM-0697-84): pages 54 (bottom) and 158; *United States in Old Maps & Prints*, by Eduard van Ermen, Lannoo, Tielt, Belgium, 1990: page 55; University of Chicago Press, for John P. Snyder, *Flattening the Earth: Two Thousand Years of Map Projections*, University of Chicago Press, 1993: pages 30 and 31 (top); University of Wales Press, Cardiff: page 108; for *The Usborne Children's Encyclopaedia*, Usborne Publishing © 1993, 1986 Usborne Publishing Ltd.: page 116; Wolverhampton Archives & Local Studies: page 16; and for *Women's Atlas of the United States*, copyright © 1995 by Timothy Fast and Cathy Carroll Fast, Facts on File, Inc., New York: page 68.

Index